Publishing Information

Get your statistics on!

homegrown statistics

Visit our website at http://mathguyzero.com

- Learn introductory statistics concepts from pie charts to chi-square tests
- Learn to solve statistical problems with a simple calculator
- Learn to use a Microsoft Excel spreadsheet as a statistics calculator
- Track your progress
- Assess your statistics skills (before your teacher does) with chapter quizzes, midterm and final exam

Homegrown Statistics is an introductory statistics textbook aligned with an online e-course from mathguyzero™. This text was developed to teach introductory statistics in a non-threatening approach to the 21st century student.

homegrown statistics

ISBN provided by:
Createspace.com

ISBN-13: 978-1463731342
ISBN-10: 1463731345

math education for the 21st century

Introduction

Welcome to 21st century math learning. Students will learn introductory statistics concepts using this textbook in conjunction with the online e-course that has over 100 video tutorials (website subscriptions sold separately at ***http://mathguyzero.com***).

Here are some helpful tips on using ***Homegrown Statistics*** to maximize the learning of statistics:

- Language is the base of learning. The ***Vocabulary Box*** illustrates the critical academic vocabulary that a student must have a solid understanding before proceeding. Learning statistics is done in stratified-learning steps, with the more complicated concepts being taught using less complicated concepts, and everything is explained through language.

Vocabulary

Categorical
variables put individuals in similar categories or groups.

Quantitative
variables can be measured with a number of units. These are also called measurements, observations and/or scores.

- The ***Professor Box*** breaks down the statistical concepts where most students struggle, and re-teaches esoteric math formulas and concepts with easier to understand math-based academic vocabulary.

Remember that a coefficient is simply a number multiplying a variable.

- The ***Computer Monkey*** graphic indicates which problems in the textbook have a corresponding online video tutorial.

- The ***MS Excel*** graphic indicates which problems in the textbook have a corresponding online Excel video tutorial. (Website subscriptions are sold separately at ***http://mathguyzero.com***).

- The ***Mathguyzero Watch the Online Video Explanations*** graphic shows the link to the guided homework videos. There is a guided homework sheet at the end of each chapter linked to online video tutorials. (Website subscriptions are sold separately at ***http://mathguyzero.com***).

- The ***Chapter Review*** graphic highlights the main concepts from the chapter

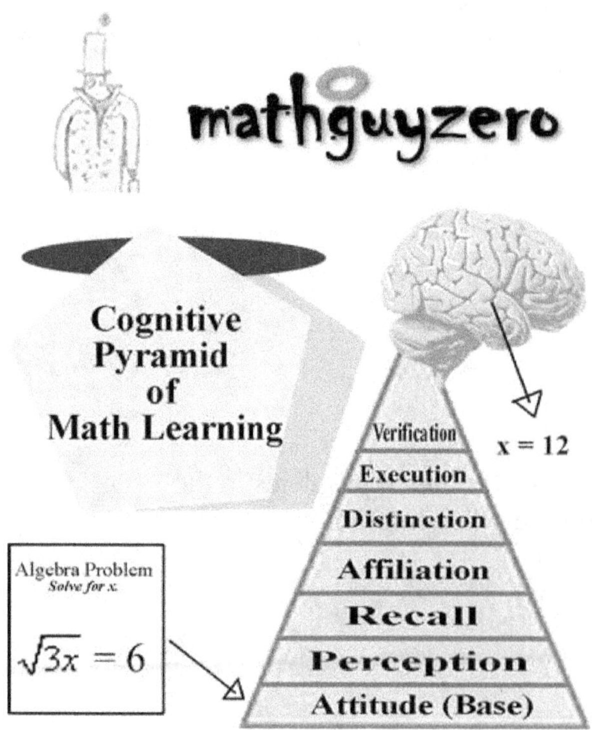

Homegrown Statistics

Table of Contents

Table of Contents

Table of Contents

Table of Contents

Table of Contents

Table of Contents

Table of Contents

Table of Contents

Table of Contents

Ch 1: Distribution Representations

Chapter

1

Distribution Representations

The truth is a pure and simple concept that cannot be altered. All statements are either true or not true. But how do we know if a statement is true? Do you believe everything that you hear from the Internet, television or read in a book? Do you trust the reports that come from these huge profit-driven companies that always state that their products work? Well, statistics helps us to decide how "true" a study is. Statistics has also been called the science of *maybe*.

The science of statistics is used for three things: to represent data, to describe data, and to make inferences from data.

In this chapter, we will investigate the basics of representing data:
- Who? - Which individuals will be studied?
- What? - What are the variables?
- How many? - Kind of self-explanatory, don't you think?
- Why? - Why are you graphing the data in the first place? - What do you want to represent?

The representation of data, or ***graphs*** of data, covered in this chapter are pie charts, bar graphs, histograms, stemplots and boxplots.

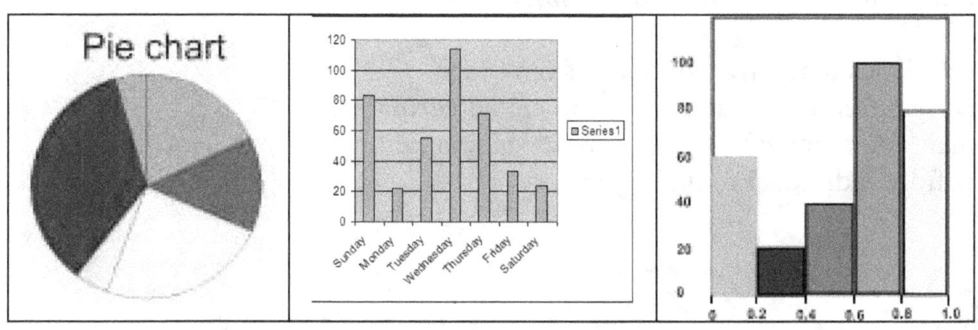

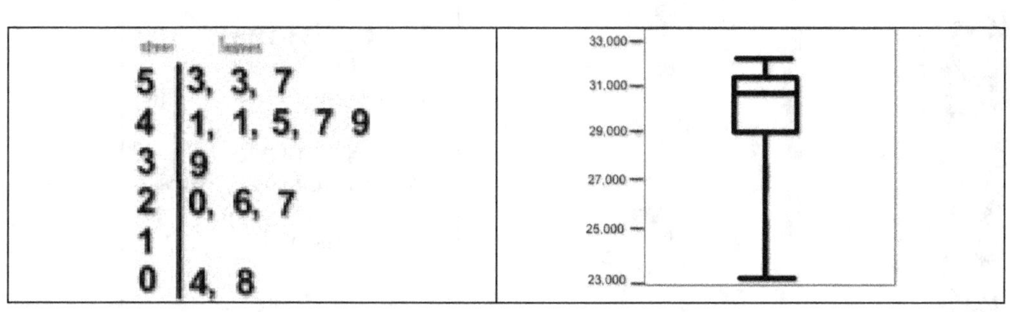

Individuals

Variables

Categorical variables
- Pie charts
- Bar graphs

Quantitative variables
- Histograms
- Stemplots
- Boxplots

Individuals and variables

Vocabulary

Descriptive
statistics ***describe*** a set of data.

Inferential
statistics make predictions, or ***inferences***, about a population from a sample.

Statistics is the science of representing, describing and making inferences about data, usually very large amounts of data.

Statistics deals with information (data) about individuals. In other words, ***individuals*** (from a group or a set) are the objects being studied. Individuals can be cats, doorknobs, avocados, test scores and/or people. When people are being studied, they are referred to as ***subjects***.

Each individual has characteristics that differ (or don't) from other individuals. These potentially different characteristics are called ***variables***. Variables can be weight, hair color, annual income or practically any characteristic that changes from individual to individual.

Vocabulary

Individuals
are the things being described by a set of data.

Variables
describe characteristics (that can change in value) of individuals.

Subjects
are human beings in a study.

Categorical variables are variables that put an individual into a specific category (hence the name). A few examples of categorical variables are: type of animal, eye color, male or female, and/or what kind of car a person drives. Notice that these characteristics are not measured with any numbered scale. ***Pie charts*** and ***bar graphs*** are used to show how many individuals are in each category.

Vocabulary

Categorical
variables put individuals in similar categories or groups.

Quantitative
variables can be measured with a number of units. These are also called measurements, observations and/or scores.

Quantitative variables are variables that are able to measure (with some kind of numbered scale) a characteristic about an individual. A few examples of quantitative variables are: age, height, weight, test scores, etc. Notice that these characteristics are measured with numbered scales, i.e., years, feet, pounds. Quantitative variables are displayed in ***histograms***, ***time plots***, ***stemplots***, ***boxplots***, and other distribution graphs.

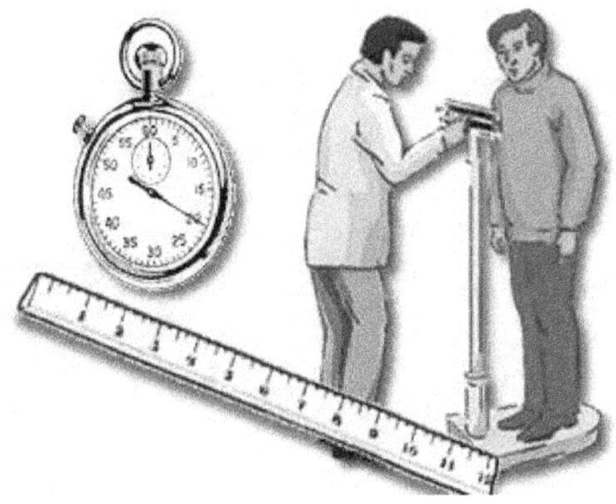

*Note: Sometimes quantitative variables can be used as categorical variables, such as people who earn between $20,000 - $30,000 can be in one category while people who earn between $50,000 - $60,000 would be in a different category. These different quantity categories are also known as ranges, class intervals or classes.

Distribution of variables

The ***distribution*** of a variable tells us how the data is arranged and how frequently each of the values occurs. Distributions show the **shape, center and spread** of the data. Imagine the distribution of books in a library. Different genres of books are in different sections of the library. The Adventure and Romance Section has over 1,000 books, the Science Fiction Section has nearly 700 books, the History Section has over 800 books while the Math Section only has a dozen or so. How sad is that?

> **Vocabulary**
>
> **Distribution** of a variable tells us how the data is spread out and how frequently each of the values occurs.

A **pie chart** is commonly used to display a distribution of a categorical variable. A pie chart is a circular chart divided into sectors with each pie "**slice**" showing the relative size of each value (of a single category) compared to the whole "**pie**" (the sum of all categories).

Instructions to make a pie chart:

Step 1) Add all of the individual categorical measurements. This sum will be the denominator to be used to find the percent of each slice.

Step 2) Divide each individual categorical measurement by the sum of all the categorical measurements and change the quotient into a percent. This percent will tell us how large each slice is compared to the entire pie.

Step 3) Draw a circle with a point in the center. All of the slices will be drawn using the circle's center as the vertex.

Step 4) Draw each slice corresponding to its percent of the whole pie.

Step 5) Label the slices, or color the slices and make a legend indicating which color represents which slice. Finished.

Example 1.1: A small library has categorized all of their books. The following table indicates how many books are in each genre.

Genre	Books
Adventure & Romance	**1032**
Science Fiction	**689**
History	**825**
Mathematics	**13**
Art	**112**
Music	**172**
Science	**84**
Government	**281**
International	**119**
Photography	**265**

Vocabulary

Pie charts illustrate how much each category is related to the sum of all categories. The "slice" represents the part and the "pie" represents the whole.

Step 1) The sum of all categorical measurements is **3592**.

Step 2)

Adventure & Romance	1032/3592	=	28.7%
Science Fiction	689/3592	=	19.2%
History	825/3592	=	23.0%
Mathematics	13/3592	=	0.4%
Art	112/3592	=	3.1%
Music	172/3592	=	4.8%
Government	281/3592	=	7.8%
International	119/3592	=	3.3%
Photography	265/3592	=	7.4%

Here is the pie chart illustrating the distribution:

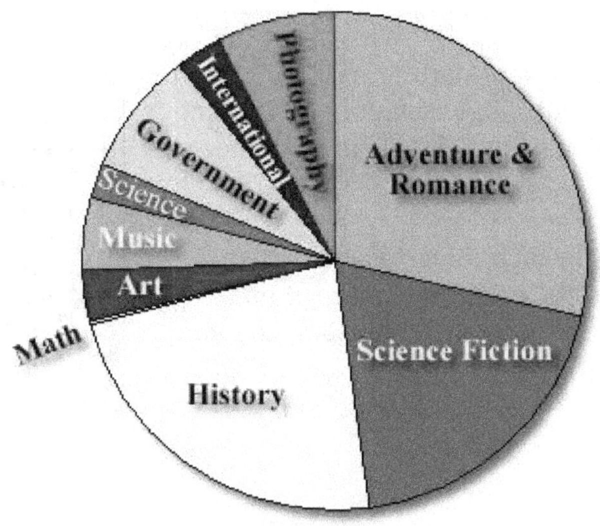

Online video example 1.2 (Lesson 1b): One hundred middle school students were asked which flavor of ice cream was their favorite and the results were:

vanilla	**22**
chocolate	**35**
strawberry	**17**
pistachio	**8**
peach	**12**
tutee fruity	**6**

Make a **pie chart** that illustrates this distribution.

Another type of distribution graph is the ***bar graph***. Bar graphs, like pie charts, show the distribution of categorical variables. The main difference between the two graphs is that pie charts are better at showing how big a part (slice) is compared to the whole (pie), while bar graphs are better at showing how each variable is related to the other variables in a distribution.

In a bar graph the different categories are lined up (in no particular order) on the horizontal (x) axis and the numerical values, or amounts of each individual in that specific category, are lined up on the vertical (y) axis beginning from least to greatest value (cardinal or ordinal).

Instructions to make a bar graph:

Step 1) Draw an x (horizontal) axis and a y (vertical) axis.

Step 2) On the x axis, make intervals of equal length with each interval containing only one category.

Step 3) On the y axis, write in a numbered scale, starting with the least (or lesser) number of individuals in any single category and ending at the greatest (or greater) number of individuals in any other single category.

Step 4) Count the number of individuals in each category and draw a rectangle with the base width as the category's interval on the x axis and a height of the number of individuals in that category.

Vocabulary

Bar graphs
illustrate how many individuals are in each category compared to the number of individuals in the other categories.

Example 1.3: A small library has categorized all of their books. The following table indicates how many books are in each genre.

Genre	Books
Adventure & Romance	**1032**
Science Fiction	**689**
History	**825**
Mathematics	**13**
Art	**112**
Music	**172**
Science	**84**
Government	**281**
International	**119**
Photography	**265**

Here is the bar graph that represents this distribution.

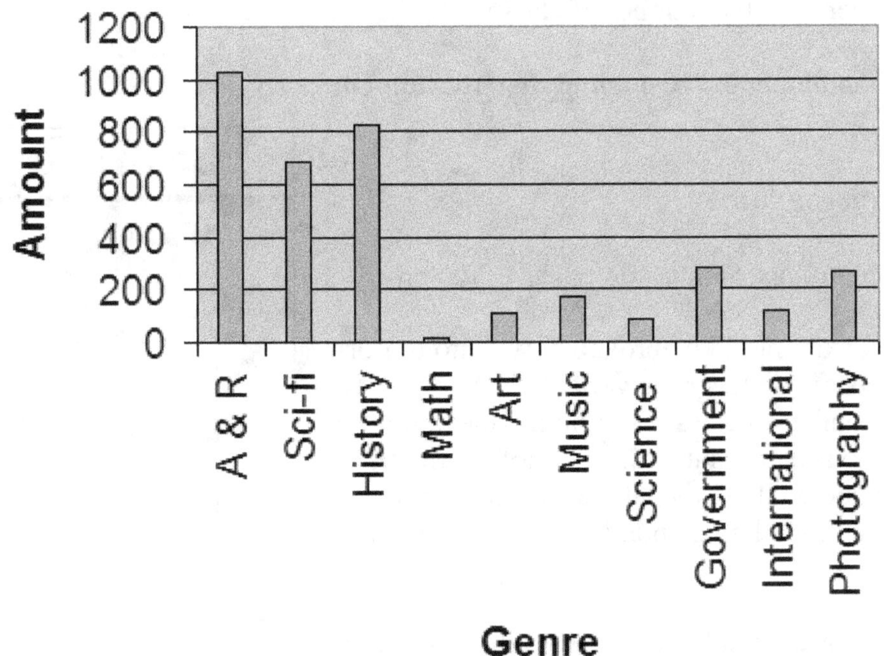

Online video example 1.4 (Lesson 1c): One hundred middle school students were asked which flavor of ice cream was their favorite and the results were:

vanilla	**22**
chocolate	**35**
strawberry	**17**
pistachio	**8**
peach	**12**
tutee fruity	**6**

Make a **bar graph** that illustrates this distribution.

A **histogram** is a type of bar graph that shows a distribution of a quantitative variable, but the horizontal axis of a histogram has ***ranges***, ***intervals***, or ***classes*** with units of measurement of whatever the variable is being measured in. The vertical axis measures the ***frequency*** of measurements in each class. Histograms are similar to stemplots in that both they show the **shape, center** and **spread** of a distribution.

The number of classes in a histogram is up to the creator of the graph, but normally there are between 5 and 15 or so classes. Class widths are determined by dividing the **range of the data** (the maximum measurement minus the minimum measurement) by the number of classes.

*Note: the histogram is the grandfather of the **normal distribution curve**, so learn it well, eh?

Vocabulary

Histograms
A histogram is a type of bar graph that shows a distribution of a quantitative variable, but the horizontal axis of a histogram has ranges of values.

Instructions to make a histogram:

Step 1) First draw an x (horizontal) axis and a y (vertical) axis.

Step 2) Label your axes: the x axis should be broken down into uniform class intervals in units of whatever the variable is being measured with; the number of classes should be relevant to the range of data. The y axis is measured off in whole numbers beginning with zero and ending at whatever whole number is equal to, or greater than, the largest number of observations in a single class interval.

Step 3) Count the number of measurements that fall into the first class range and draw a rectangle with the height being the number of measurements and the width of a single class interval.

Step 4) Repeat step #3 until all of the measurements have been placed into a class interval. Done and done.

Example 1.5: 320 college students had a GPA of 1.0 or less. Here is the distribution of their GPAs.

GPA	Number
0.0 - 0.2	**60**
0.2 - 0.4	**20**
0.4 - 0.6	**100**
0.6 - 0.8	**80**
0.8 - 1.0	**60**

Step 1) First draw an x (horizontal) axis and a y (vertical) axis that "fit" this problem: the x axis would be the GPA categories; the y axis would be the frequency, or the number of individuals who are in each category.

Step 2) Start with the first category: there are 60 individuals in the category of 0.0 – 0.2 GPA. Make a rectangle with a base length of 0.2 and a height of 60 units.

Step 3) Repeat the process for all categories. Done and done.

Here is the histogram of this distribution:

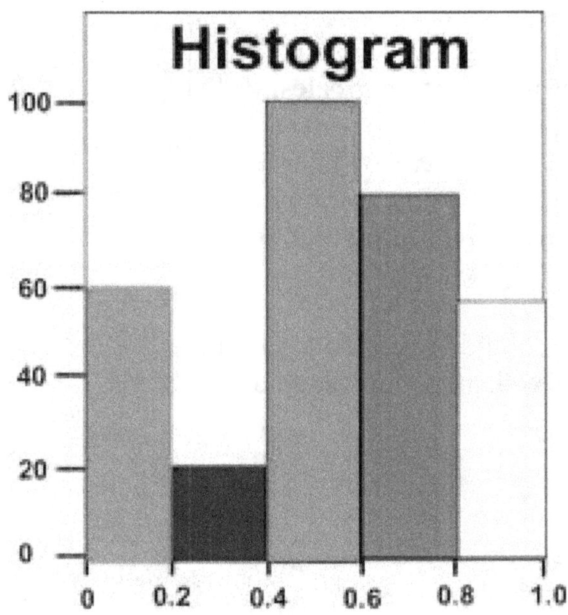

*Note: the class ranges are equal to, or less than, the largest value in that category. In other words, if the variable were exactly 0.2 it would be placed in the first column because 0.2 is equal to the largest of that class value of 0.2.

Online video example 1.6 (Lesson 1d): Seventeen households were polled about their household annual incomes. Below were the responses in US dollars. Make a **histogram** of the data.

16438	71324	46770
36409	18968	19006
55933	42558	50770
34881	89688	24885
23039	64985	43980
28966	90644	

A **stemplot** (or a stem-and-leaf plot) is a representation of quantitative data in an easy to interpret format, similar to a histogram, in order to assist in visualizing the shape and spread of a small distribution.

A basic stemplot contains two columns of numbers separated by a vertical line. The left column contains the ***stems*** and the right column contains the ***leaves***. Stems represent the leading digits that many of the measurements share. The leaves are the ending digits of the individual measurements.

*Note: if you rotated a stemplot 90 degrees counterclockwise it would function (and look like) a histogram.

Vocabulary

Stemplots are graphs that show the shape and spread of a distribution of a quantitative variable using shared leading digits of units of measurements.

Instructions to make a stemplot:

Step 1) List all of the measurements in order from least to greatest.

Step 2) Break apart all measurements into *stems* and *leaves*. The stems are the shared leading digit, or digits, (the number or numbers that are the left-most) and the leaves are the ending digit, or digits, (the number or numbers that are the right-most).

(*Note: the leading digit or digits, the stems, should be related to all of the measurements in such a way that they display the data in a functional and uncomplicated manner.)

Step 3) Draw a vertical line long enough to write all of the *stems* on the left side, from the smallest stem at the bottom to the largest stem at the top.

Step 4) Line up all of the *stems* vertically along the vertical line with the smallest at the bottom and the largest at the top.

Step 5) Line up all of the leaves to their corresponding stems vertically and horizontally on the right side of the vertical line. Finished.

Example 1.7: Make a stemplot from the following data: 29143, 33565, 34084, 30449, 29874, 31843, 33739, 32267, 30704, 33045, 31543, and 31838.

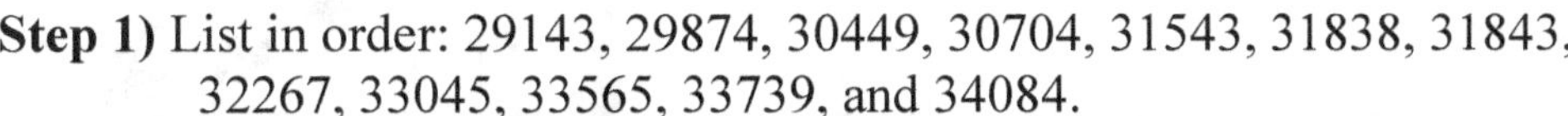

Step 1) List in order: 29143, 29874, 30449, 30704, 31543, 31838, 31843, 32267, 33045, 33565, 33739, and 34084.

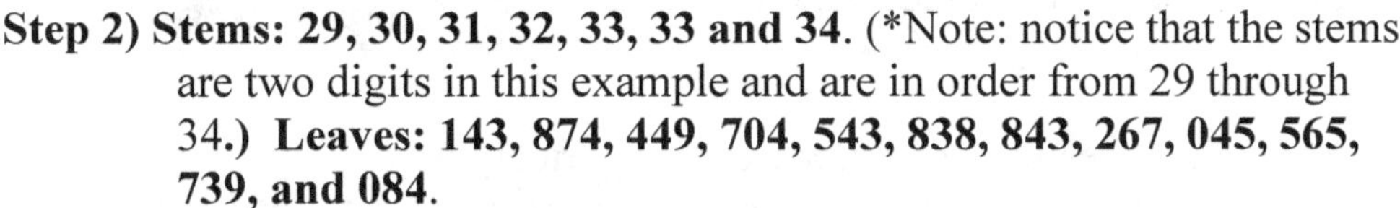

Step 2) Stems: 29, 30, 31, 32, 33, 33 and 34. (*Note: notice that the stems are two digits in this example and are in order from 29 through 34.) **Leaves: 143, 874, 449, 704, 543, 838, 843, 267, 045, 565, 739, and 084**.

Steps 3, 4 & 5)

Stem	Leaves
34	084
33	045, 565, 739
32	267
31	543, 838, 843
30	449, 704
29	143, 874

Online video example 1.8 (Lesson 1e): Make a **stemplot** out of the following measurements.

4	**41**
8	**45**
20	**47**
26	**49**
27	**53**
39	**53**
41	**57**

Chapter 1 Review

A set of data contains information about ***individuals***. Certain characteristics about individuals can be given values which we call ***variables***.

Some variables are ***categorical*** while others are ***quantitative***. Categorical variables place individuals into specific categories. Quantitative variables are measurements about a characteristic of an individual with a specific number of units.

The ***distribution*** of a variable shows the **shape, center and spread** of the data and illustrates what values the variable takes on, plus how frequently each value occurs.

Categorical variables are represented in ***pie charts*** (how one category is related to the sum of all categories) and in ***bar graphs*** (how one category is related to the other categories).

Quantitative variables are represented by ***histograms*** (a bar graph but with equal ranges of values on the x axis) and by ***stemplots***.

Guided homework Chapter 1: Distribution Representations

1) The following individual variables were studied in an insurance company report: gender, alcohol user, ethnicity, smoker, height, weight, age, income, distance driven daily, and blood pressure. List the categorical variables and the quantitative variables.

2) Jake works very hard but he seems to constantly run out of money before the end of the month. He wants to know where his money goes so he can make a budget. Here are his monthly expenditures: rent - $650, food - $200, phone - $80, car payment - $300, car insurance - $75, gasoline - $120, credit cards - $250, cable - $100, gas & electric - $140, water - $80, clothing - $100 and entertainment - $150. Draw the pie chart representing where Jake's money goes.

3) A small local car dealership was recently dumped by General Motors. They had to sell their entire inventory. The general manager made a list of the colors of the cars that needed to be sold: 26 silver, 13 white, 14 blue, 9 green, 4 brown, 19 black, and 12 red. Make a bar chart to represent the number of cars by color.

4) The following are the selling prices of single family homes in San Diego, California in 2007. Make a histogram of the data. (In thousands of dollars.)

504.3	**498.1**
468.7	**604.8**
339.7	**582.5**
518	**519.4**
628.3	**683.4**
751.2	**705.7**
544.9	**391.1**
480.6	**508.6**

5) The following data represents test scores from Mr. Regan's algebra class. Make a stemplot of their scores.

76	100	93
0	70	22
67	92	41
76	82	68
42	82	70
31	100	74
	15	79

6) Convert the stemplot below into a list of standard data.

5	3, 3, 7
4	1, 1, 5, 7 9
3	9
2	0, 6, 7
1	
0	4, 8

Chapter 1 Quiz

1. Fill in the blank with the correct word.

Anything being studied in a statistical study is called an "individual", but when that individual is a human being, we call them "_______".

a. people
b. researchees
c. subjects
d. objects
e. human guinea pigs

2. True or false?

A person's age is a **quantitative** variable.

3. True or false?

A person's blood type is a **quantitative** variable.

4. True or false?

Ranges, or intervals, of quantitative variables can be used as categorical variables.

5. According to the pie chart, which day of the week is the busiest day for bandwidth?

a. Monday
b. Saturday
c. Wednesday
d. Friday
e. Sunday

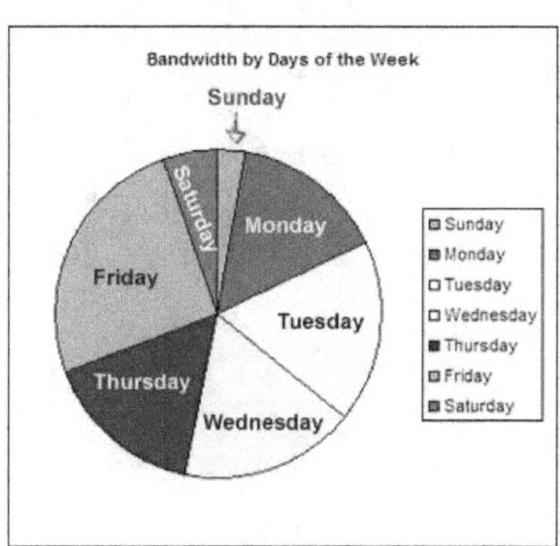

Pie chart

6. According to the pie chart, which day of the week is the slowest day for bandwidth?

a. Monday
b. Saturday
c. Wednesday
d. Friday
e. Sunday

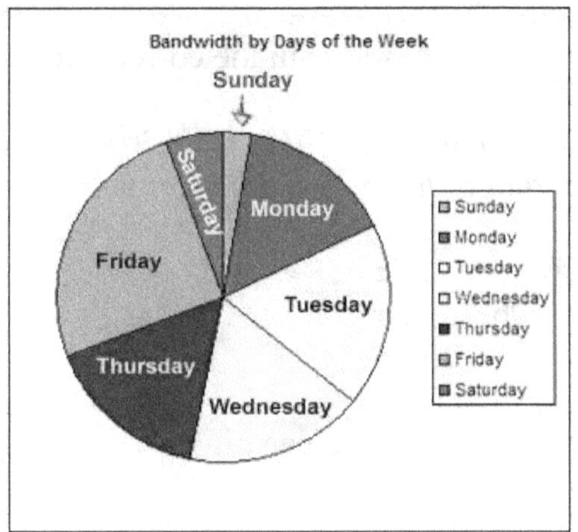

7. Which bar chart below matches the data of the shown pie chart above?

a.

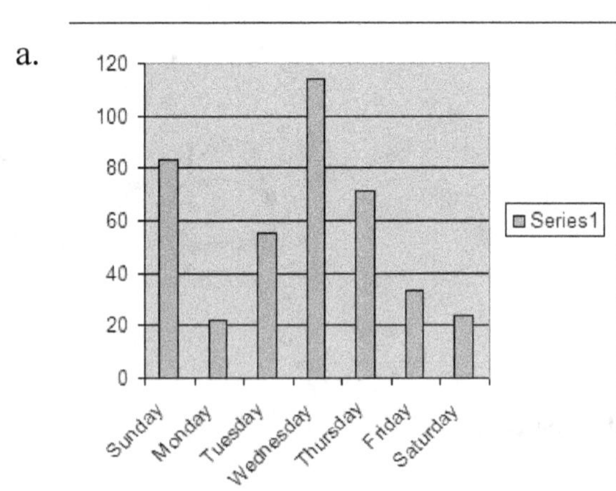

b.

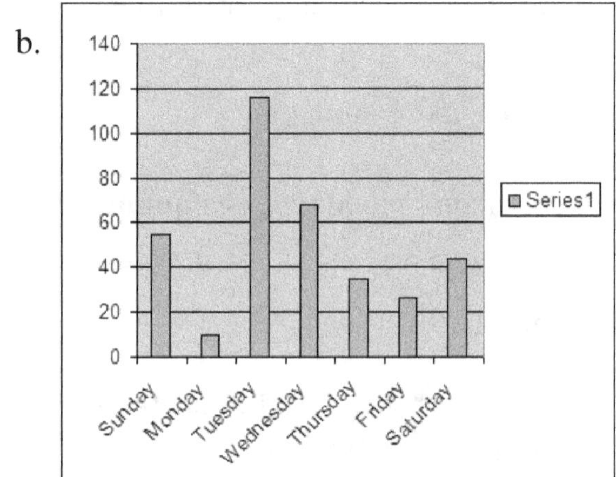

c.

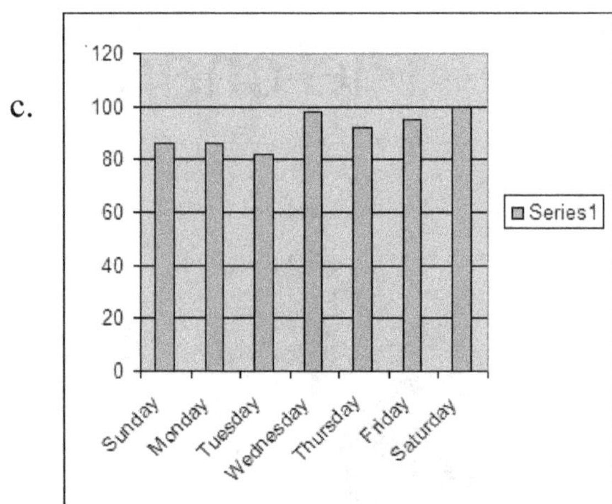

d.

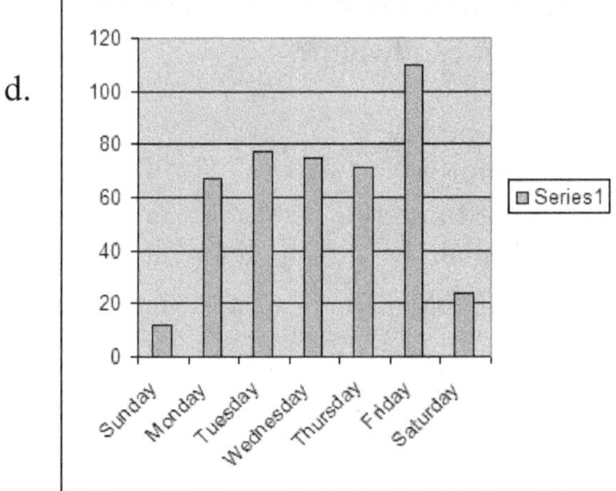

8. Which of the following ranges, or intervals, would best fit a histogram with 6 categories with the data below?

24318	35906
30620	23425
36638	24108
37359	31478
25400	22770
33180	14360
28620	32770
27440	26117
22472	35210
21033	10276

a. 6 ranges of 10000
b. 6 ranges of 1000
c. 6 ranges of 5000
d. 6 ranges of 20000
e. 6 ranges of 3000

9. According to the histogram on the right, how many students have a GPA between 3.0 and 4.0?

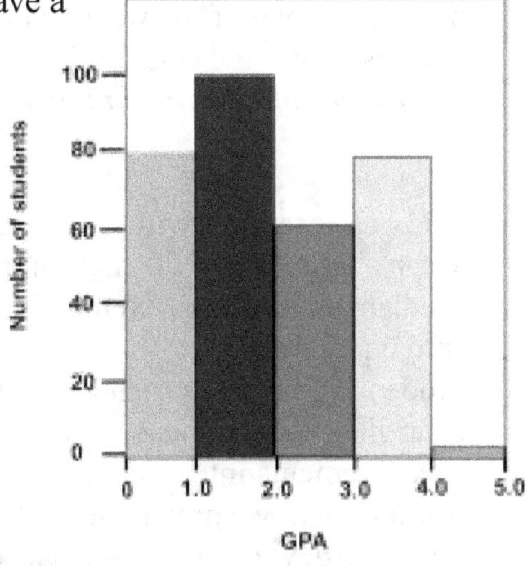

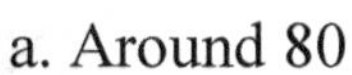

a. Around 80
b. Around 100
c. Around 60
d. Around 20
e. Around 160

10. Which of the following is not on the stemplot?

a. 109
b. 19
c. 87
d. 69
e. 71

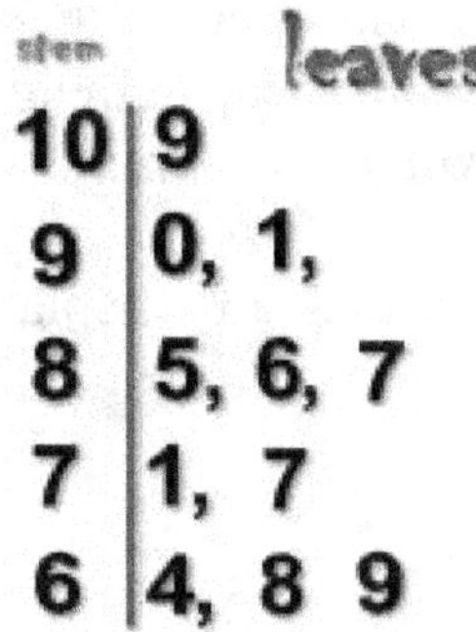

Ch 2: Measures of Central Tendencies

Chapter 2

Measures of Central Tendencies

After data has been collected and a graphical distribution of the data has been made, what next? There are many different things to look for, but exactly what depends on the researcher. But without a doubt, the absolute king of statistics is the ***mean***, or the average, of the data. The mean is used in nearly every quantitative statistical study.

The mean is one of the ways to *describe* a set of measurements. It indicates where the center of the measurements tend to be, thus the name of ***measures of central tendencies***.

In this chapter, we will investigate the measures of central tendencies:
- Mean - are most of the measurements close to the mean, or not?
- Median - which measurement, or observation, divides the entire group into equal halves?
- Mode - which measurement is repeated the most frequently?
- Quartiles - which measurements, or observations, divide the set of measurements into quarters?
- Variance - how spread out are the measurements?
- Standard deviation - what is the average distance away from the mean?
- Outliers - what do you do with measurements that aren't like the others?
- Five-number summary - which measurements are best to use to describe the data?
- Boxplot - how would you graphically display your findings?

Mean, Median and Mode

Variance

Standard Deviation

Quartiles and Outliers

Five-number Summary

Boxplot

Mean, median and mode

The **mean** ($\bar{X}$ or μ), or average, normally refers to an arithmetic mean. To find an arithmetic mean, add up all of the measurements and divide that sum by the number (***n***) of measurements. The mean is the most used math concept in statistics, so learn it well! ***Means can be seriously affected by outliers***.

$$\bar{X} = \frac{\sum x_i}{n}$$

$\bar{X}$ is the symbol for the mean.

$\sum$ is the symbol for "add them all up".

X_i is the symbol for "each and every" measurement.

— is the symbol for division.

n is the number of measurements.

Here is the formula to find the (arithmetic) mean:

$$\bar{X} = \frac{\sum x_i}{n}$$

Don’t let the capital sigma scare you. It simply means *summation of*, or *just add them all together*.

Here is another mean formula that is easier to read:

$$\bar{X} = \frac{X_1 + X_2 + X_3 + \ldots + X_n}{n}$$

Example 2.1: Find the mean of the following test scores (measurements): 45, 56, 59, 38, 48, 42, 61 and 55.

Step 1) Add up all of the measurements: 45 + 56 + 59 + 38 + 48 + 42 + 61 + 55 = 404

Step 2) Divide the sum (404) by the number of measurements (represented by the variable "***n***" – the number of test scores: in this example n = 8). So, 404 divided by 8 equals a **mean of 50.5**.

*Warning: We must be careful about the mean! Here is an example as to why: let's say you put one foot in a bucket of water and then froze it to -100 degrees Fahrenheit, and your other foot in a bucket of boiling water at 212 degrees F. The average temperature is (-100 + 212) divided by 2 = 56 degrees. Notice that the mean temperature would be a bit chilly to swim in, but in reality both of your feet would be... gone… painfully so!

The **median** is the measurement, or observation, separating the measurements in half, with 50% of the measurements greater than the median and 50% less than the median. **If there is an even number of observations, the median is the mean of the two middle values**. At most, half the population has values less than the median and, at most, half have values greater than the median. ***Medians are not seriously affected by outliers.***

*Note: the median is one of the measurements in the data set, or the average of the two measurements, which is in the exact middle of all of the ordered measurements.

Example 2.2: Find the median of the following test scores: 75, 82, 62, 55, 95, 99, 76, 82 and 81.

Step 1) List in order: 55, 62, 75, 76, 81, 82, 82, 95 and 99.

Step 2) Find the measurement that is exactly in the center: 81 (Notice that 4 measurements are less than 81 and 4 measurements are greater that 81.)

Step 3) If there is an even number of measurements in your data set, then there will be two measurements in the middle. When this happens use the average of the two middle measurements as the median.

Greek and English variables

Populations have ***parameters*** which are represented by Greek letters.

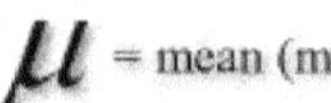 = mean (mu)

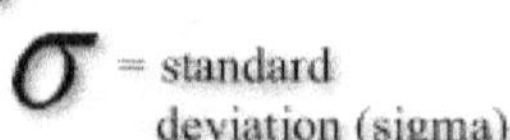 = standard deviation (sigma)

Samples have ***statistics*** which are represented by English letters.

 = mean (x-bar)

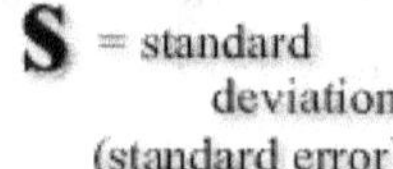 = standard deviation (standard error)

The **mode** is the value that occurs the most frequently in a data set or a probability distribution. Like the statistical mean and the median, the mode is a way of capturing important information about a random variable or a population in a single quantity. There can be more than one mode in a set of measurements, or there can be no modes at all!

Example 2.3: Find the mode of the following test scores: 75, 82, 62, 55, 95, 99, 76, 82 and 81.

Step 1) Put all of the measurements in order: 55, 62, 75, 76, 81, 82, 82, 95 and 99.

Step 2) Find the measurement that is the most frequent: 82.

Note: the values of the ***mean*, ***median*** and ***mode*** from a ***normal distribution*** are relatively close to each other.

Vocabulary

Mean
The mean is the arithmetic average of a set of data. Add all measurements and then divide that sum by the number of measurements in the data set.

Median
The median is the measurement exactly in the middle of a set of data. If the data set has two measurments in the middle, add them together and divide by two to find the median.

Mode
The mode is the most frequent measurement in a data set. There can be more than one mode.

Variance

Variance is a form of the verb "*to vary*", which means to differ. The variance of a data set defines how the measurements are spread out. A large variance indicates that the data is spread far apart, while a small variance indicates that most measurements are close to the mean. A variance from a sample data set is represented with s^2, and from an entire population as σ^2 (sigma squared).

There are four steps to find the variance of a data set.

Step 1) Find the mean of the data set.

Step 2) Subtract each measurement from the mean and then square that difference.

Step 3) Add up each and every squared difference from step 3.

Step 4) ***IF THE DATA SET IS A POPULATION***, divide the sum from step 3 by the number of measurements in the data set; ***IF THE DATA SET IS A SAMPLE***, divide the sum from step 3 by ***1 less than the number*** of measurements in the data set.

Here is the formula to find the variance of a population:

$$\sigma^2 = \frac{\Sigma(\mu - x_i)^2}{n}$$

variance formula for a POPULATION

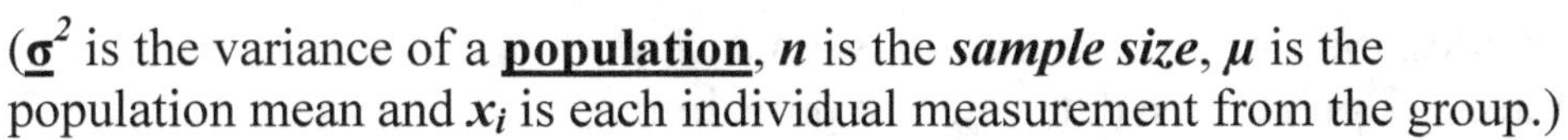

(σ^2 is the variance of a **population**, ***n*** is the ***sample size***, μ is the population mean and x_i is each individual measurement from the group.)

$$s^2 = \frac{\Sigma(\bar{X} - x_i)^2}{n-1}$$

variance formula for a SAMPLE

(s^2 is the variance of a **sample**, ***n*** is the sample size, ***x-bar*** is the sample mean and x_i is each individual measurement from the group.)

Vocabulary

Population
A population is ALL individuals in a set. Example: the population of chickens is every chicken in the world. That's a lot of chickens!

Sample
A sample can be any number of individuals from a set. Example: Three chickens.

Greek and English variables

Populations have ***parameters*** which are represented by Greek letters.

μ = mean (mu)

σ = standard deviation (sigma)

Samples have ***statistics*** which are represented by English letters.

$\bar{X}$ = mean (x-bar)

S = standard deviation (standard error)

Example 2.4: Find the variance from the following **sample** data: 75, 82, 62, 55, 95, 99, 76, 82 and 81.

Step 1) Find the mean: (75 + 82 + 62 + 55 + 95 + 99 + 76 + 82 + 81) divided by 9 = 78.56.

Step 2) Subtract the group mean from each and every individual measurement, one at a time. (75 - 78.56 = -3.56), (82 - 78.56 = 3.56), (62- 78.56 = -16.56), (55 - 78.56 = -23.56), (95 - 78.56 = 16.44), (99 - 78.56 = 20.44), (76 - 78.56 = -2.56), (82 - 78.56 = 3.44) and (81 - 78.56 = 2.44).

Step 3) Square each of the differences: $(3.56)^2 = 12.64$, $(3.44)^2 = 11.86$, $(-16.56)^2 = 274.42$, $(-23.56)^2 = 554.86$, $(16.44)^2 = 270.42$, $(-2.56)^2 = 6.53$, $(3.44)^2 = 11.86$ and $(2.44)^2 = 5.98$.

Step 4) Add up all of the squared differences: 12.64 + 11.86 + 274.09 + 554.86 + 270.42 + 417.98 + 6.53 + 11.86 + 5.98 = 1566.22.

Step 5) Sum up all squared differences and divide by 1 less than the **sample** size: 1566.22 divided by (9 - 1) = **195.78**. This final calculation gives the ***variance***.

*Note: several different data sets can all have the same mean, but their variances can be extremely different.

variance formula

(population)

$$\sigma^2 = \frac{\Sigma(\mu - x_i)^2}{n}$$

σ^2 is the symbol for the **population** variance.

(sample)

$$s^2 = \frac{\Sigma(\bar{X} - x_i)^2}{n-1}$$

s^2 is the symbol for the **sample** variance.

Standard deviation

In probability theory and statistics, the ***standard deviation*** is a measure of the variability, or dispersion, of a population, a data set, or a probability distribution. A small standard deviation indicates that the data points tend to be very close to the same value (the mean), while a large standard deviation indicates that the data are "spread out" over a large range of values.

Another way to think about a standard deviation is that it is the average distance of each individual measurement away from the group's mean, or more succinctly, the mean distance away from the mean. Cool!

*Note: **The standard deviation squared is equal to the variance**. And that means that the **square root of the variance is equal to the standard deviation**.

Note: The standard deviation is represented by the **English letter ***s*** when it is from **sample** and the **Greek** letter sigma ***(σ)*** when it is from a *population.*

Here is the standard deviation formula for a **population**:

$$\sigma = \sqrt{\frac{\Sigma(\mu - x_i)^2}{n}}$$

(***σ*** is the symbol for the population standard deviation, ***n*** is the sample size, x_i represents each individual measurement and ***μ*** represents the population mean.)

Below is the standard deviation formula for a **sample**:

$$s = \sqrt{\frac{\Sigma(\bar{X} - x_i)^2}{n-1}}$$

(***s*** is the symbol for the population standard deviation, ***n*** is the sample size, x_i represents each individual measurement and ***x-bar*** represents the sample mean. The standard deviation from a sample is called a **standard error**.)

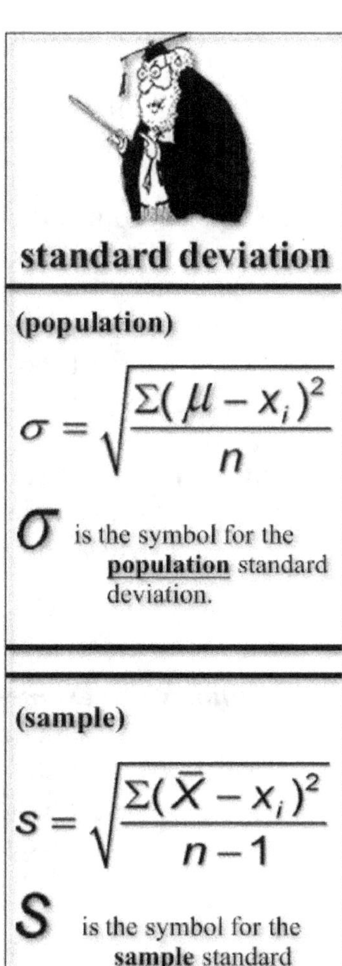

Example 2.5: Find the standard deviation from the following sample test scores: 75, 82, 62, 55, 95, 99, 76, 82 and 81.

Step 1) Find the mean: (75 + 82 + 62 + 55 + 95 + 99 + 76 + 82 + 81) divided by 9 = 78.56

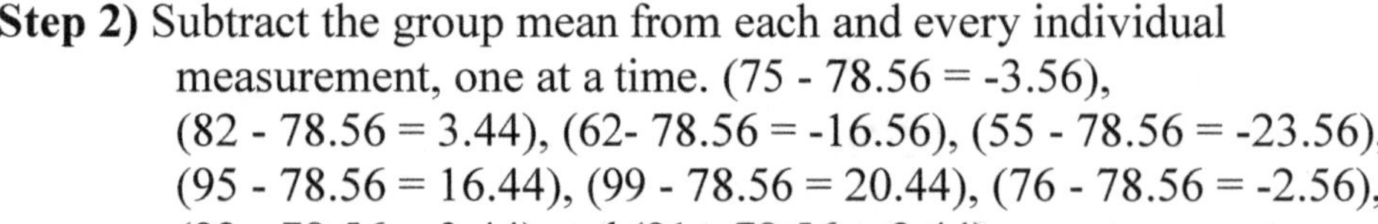

Step 2) Subtract the group mean from each and every individual measurement, one at a time. (75 - 78.56 = -3.56), (82 - 78.56 = 3.44), (62- 78.56 = -16.56), (55 - 78.56 = -23.56), (95 - 78.56 = 16.44), (99 - 78.56 = 20.44), (76 - 78.56 = -2.56), (82 - 78.56 = 3.44) and (81 - 78.56 = 2.44).

Step 3) Square each of the differences: $(3.56)^2 = 12.67$, $(3.44)^2 = 11.83$, $(-16.56)^2 = 274.23$, $(-23.56)^2 = 555.07$, $(16.44)^2 = 270.27$, $(20.44)^2 = 417.79$, $(-2.56)^2 = 6.55$, $(3.44)^2 = 11.83$ and $(2.44)^2 = 5.95$.

Step 4) Add up all of the squared differences: 12.67 + 11.83 + 274.23 + 555.07 + 270.27 + 417.79 + 6.55 + 11.83 + 5.95 = 1566.22.

Step 5) Divide the sum of all squared differences by 1 less than the sample size: 1566.22 divided by (9 - 1) = 195.78. (This is the ***sample*** variance, don't you know?)

Step 6) Take the square root of the variance: the square root of 195.78 = 13.99. (This is the standard deviation of the sample data – we will later call the *sample standard deviation* the ***standard error***.)

Online video example 2.6: A giant faceless money-hungry pharmaceutical company is conducting a research study with 478 subjects being treated for high systolic blood pressure. All subjects were given a new treatment medication called Calm-it-all. Six subjects were selected at random and their blood pressures were taken and recorded. Here are their systolic measurements:

126, 131, 118, 141, 139 and 131

Find the mean, median, mode, variance and standard deviation of the blood pressures.

Quartiles

In descriptive statistics, a ***quartile*** (**Q**) is a specific measurement from a data set which divides the ranked, or sorted data, into four equal parts, so that each quartile represents one fourth of all of the measurements from a population sample.

Vocabulary

Quartile
Quartiles are individual measurements in a data set that separate the measurements into four groups with the same number of measurements in each group.

*Note: there is another divider of data called a *decile*. It works just like a quartile, but it separates the measurements into 10 different groups instead of four. The prefix of "dec-" means 10, don't you know? But why is December the twelfth month? Mgz knows, do you?

Example 2.7: Find the first and third quartile from the following data:
75, 82, 62, 55, 95, 99, 76, 82 and 81.

Step 1) Put the data in order from least to greatest value:
55, 62, 75, 76, 81, 82, 82, 95, 99.

Step 2) Since there are 9 measurements, divide 9 by 4 = 2 1/4, and then round up to the nearest whole number. In this case, 2 1/4 rounds up to 3. Does the 3rd measurement have 25% (inclusively) of the data below it? (Why, yes it does!) So **q1 = 75**.

Step 3) Since there are 9 measurements, divide 9 by 4 = 2 1/4, and then multiply that by **3** (because it is the **third** quartile) equals 6 3/4 which rounds up to 7. Does the 7th measurement have 75% of the data below it (inclusively)... and 25% of all measurement above it? (Why, yes it does!) So **q3 = 82**.

*Note: The *second quartile* is the same as the *median* and the **fourth quartile** is the same as the **maximum**.

Outliers

An ***outlier*** is any measurement from a statistical study that is either much smaller than, or much greater than, most of the other measurements. Outliers can strongly affect the results from certain statistical tests. How much greater, or smaller, a measurement has to be in order for it to be labeled as an outlier is a matter for discourse between statisticians everywhere. But here are a couple of largely accepted outlier identifiers.

Vocabulary

Outlier
An outlier is a measurement in a data set that is far greater or far smaller than most of the other measurements.

One rule to identify an outlier is if the suspected measurement is more than, or less than, three standard deviations away from the mean of the group, because the ***Empirical Rule*** states that 99.7% of all measurements should fall within the range of the mean plus or minus three standard deviations.

Another rule to identify an outlier is to use the ***Interquartile Rule***, which states that any measurement might be an outlier if it is 1.5 times the IQR below quartile 1, or above quartile 3. The IQR (interquartile range) is equal to **Q3 - Q1**.

*Note: it is up to the researchers to defend their reasons behind removing a measurement from a group of measurements, because after all, the suspected outlier is valid data and may mean something, maybe something important.

> **Vocabulary**
>
> **Empirical Rule**
> The Empirical Rule states that 68% of all measurements will fall between the group mean plus/minus one standard deviation, 95% will fall between the mean plus/minus two standard deviations, and 99.7% will fall between the mean plus/minus three standard deviations

Example 2.8: Identify any and all outliers from the following data:
26, 16, 24, 22, 119, 21, 25, 23, 24, 22 and 26.

(**Empirical Rule method**)
Step 1) Find the mean and the standard deviation of the data:
x-bar = 31.64 and **s = 29.11**.

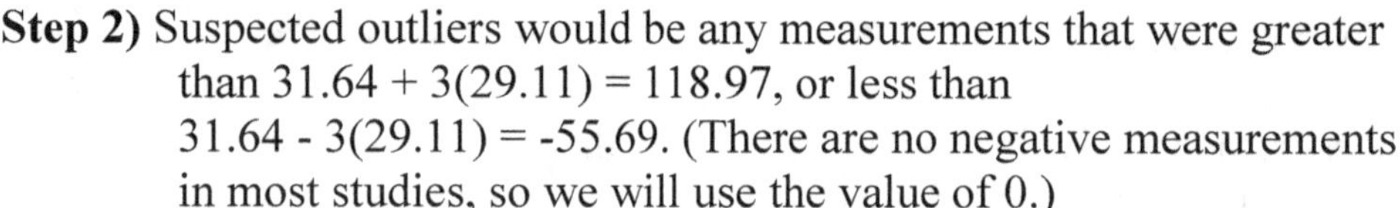

Step 2) Suspected outliers would be any measurements that were greater than 31.64 + 3(29.11) = 118.97, or less than 31.64 - 3(29.11) = -55.69. (There are no negative measurements in most studies, so we will use the value of 0.)

Step 3) The only outlier identified with the empirical method was **119**.

> **Vocabulary**
>
> **IQR Rule**
> The Interquartile Rule idenitfies possible outliers. If a measurement is more than 1.5(q3 - q1) above the third quartile, or less than 1.5(q3 - q1) below the first quartile, then it may be considered an outlier.

(**IQR rule**)
Step 1) Find the interquartile range (IQR): Q3 = 25.5 and Q1 = 22, so Q3 - Q1 = 25.5 - 22 = **3.5**.

Step 2) Multiply the IQR by 1.5: **3.5 x 1.5 = 5.25**

Step 3) Suspected outliers would be any measurements that were greater than Q3 by at least 5.25, or less than Q1 by at least 5.25.

Step 4) The outliers identified with the IQR rule were **16** and **119.**

**Note: sometimes outliers can be caused by faulty data collecting. Always double check the accuracy of your data. It couldn't hurt!

Five-number summary

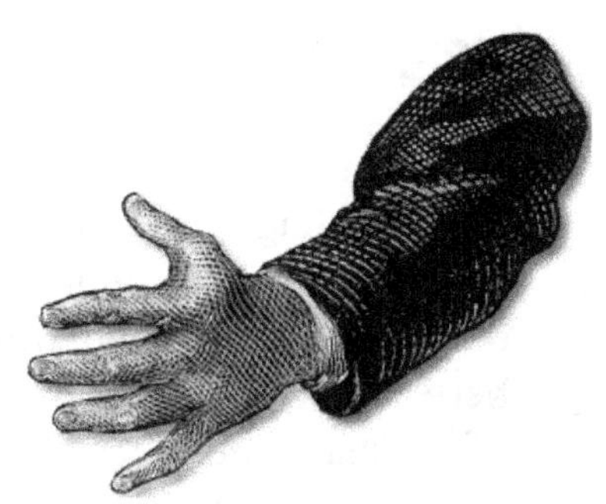

The ***five-number summary*** illustrates the variation, or the spread, or the range of a group of data. It contains the following information about a group of measurements or observations: **minimum** (the smallest measurement), **first quartile**, **median**, **third quartile** and the **maximum** (the largest measurement) (min, Q1, median, Q3, max). These five data points are used to draw a *boxplot* which is an easy-to-read graph of these data points.

The five-number summary (and the boxplot) is usually a better way to illustrate the spread of a skewed distribution, or a distribution with outliers.

Example 2.9 : Find the five-number summary from the following data:
75,82, 62, 55, 95, 99, 76, 82 and 81.

Step 1) Put the data in order from least to greatest value.
55, 62, 75, 76, 81, 82, 82, 95, 99.

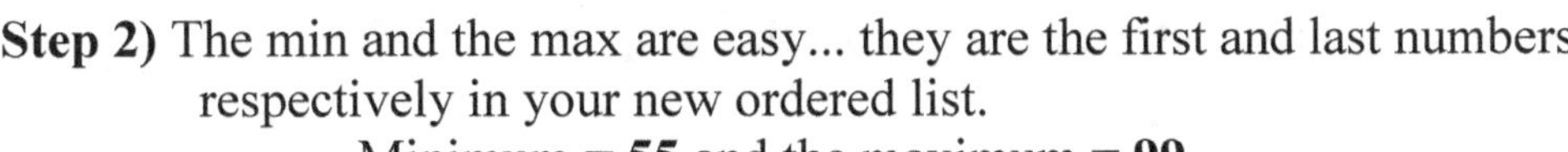

Step 2) The min and the max are easy... they are the first and last numbers respectively in your new ordered list.
Minimum = **55** and the maximum = **99**.

Step 3) The median of the data is that measurement which has half of the measurements above it and half of the measurements below it.
Median = **81**.

Step 4) The first quartile (Q1) is the measurement in the set of data that has 1/4th of the other measurements below it. Q1 = **75**.

Step 5) The third quartile (Q3) is the measurement in the set of data that has 3/4ths of measurements below it (and 1/4th of the data above it). Q3 = **82**.

Step 6) Write the terms out in this order: (min, Q1, median, Q3, max).
So the **five-number summary is (55, 75, 81, 82, 99)**.

> **Vocabulary**
>
> **Five-number Summary**
> The five-number summary is a set of measurements from a data set that illustrates the spread of the data using the minimum, quartile 1, median, quartile 3 and the maximum from a set of data.

Remember that quartiles divide the measurements from a data set into groups of 4, quintiles into groups of 5, deciles into groups of 10… get the picture?

Boxplots

In descriptive statistics, a ***boxplot*** or box plot (also known as a box-and-whisker diagram or plot) is a good way of graphically displaying groups of numerical data through their five-number summaries (minimum, first quartile (Q1), median, third quartile (Q3), and maximum). The spaces between the different parts of the box illustrate the spread and/or skewness of the data, and identify outliers. Boxplots can be drawn either horizontally or vertically.

The boxplot (and the five-number summary) is usually a better way to illustrate the spread of a skewed distribution, or a distribution with outliers.

Instructions to make a boxplot:

Step 1) Find the five-number summary (minimum, first quartile, median, third quartile and the maximum) from all of the observations.

Step 2) Draw a horizontal and a vertical axis. (*Note: boxplots can go either horizontally or vertically. In the next example the boxplot will be horizontal.)

Step 3) Label the vertical axis with uniform intervals beginning with a value equal to, or less than, the minimum at the bottom, and a value equal to, or greater, than the maximum at the top.

Step 4) Draw a short *horizontal* line at each of the data points from the five-number summary.

Step 5) Draw two short *vertical* lines (to make a "box") from the first and third quartiles. Finished.

Vocabulary

Boxplot
The boxplot uses the measurements from a five-number summary of a set of data that illustrates the spread of the data using the minimum, quartile 1, median, quartile 3 and the maximum from a set of data.

Example 2.10: Make a boxplot from the following data: 75, 82, 62, 55, 95, 99, 76, 82, 81, 55, 62, 75, 76, 81, 82, 82, 95 and 99.

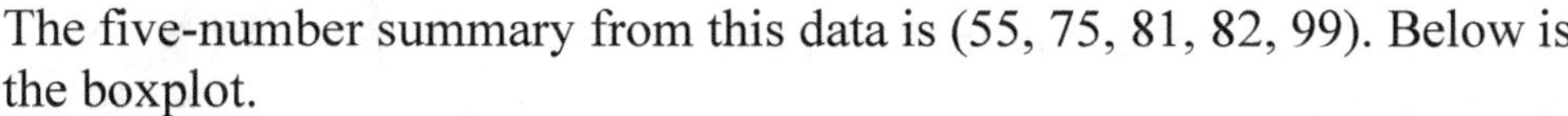

The five-number summary from this data is (55, 75, 81, 82, 99). Below is the boxplot.

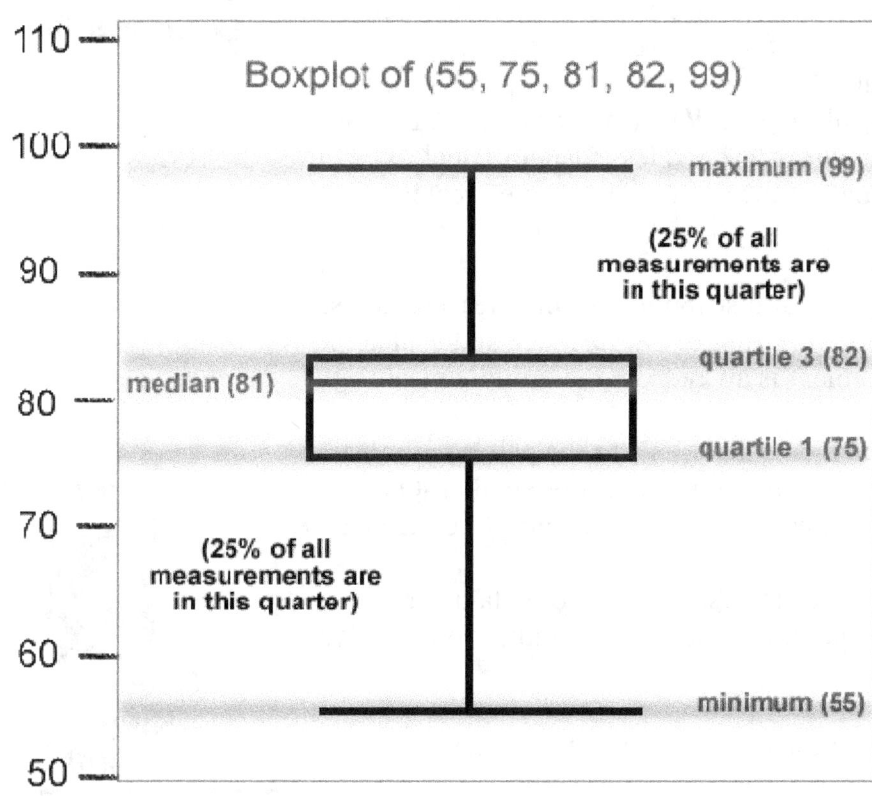

Online video example 2.11: The following are test scores from a statistics class. Make a five-number summary and a boxplot from this data. (Does your teacher do this?)

75, 82, 62, 55, 95, 99, 76, 82, 81, 55, 62, 75, 76, 81, 82, 82, 95, and 99.

Chapter 2 Review

Measures of central tendencies describe a set of data. The ***mean*** is the average of a data set, the ***median*** is the measurement in the middle and the ***mode*** is the most frequent measurement.

Quartiles are those measurements that divide the set of measurements into fourths. In other words, 25% of all of the measurements are equal to, or less than, quartile 1; 25% of all of the measurements are greater than, ***quartile 1*** and less than, or equal to, ***quartile 2*** (median); 25% of all of the measurements are greater than ***quartile 2*** and less than, or equal to, ***quartile 3***; and finally 25% of all of the measurements are greater than ***quartile 3***.

The ***variance*** indicates the spread, or distribution, of the measurements. The ***standard deviation*** is the *square root of the variance* and is also the average distance each measurement is away from the group's mean.

Outliers are individual measurements that don't fit the pattern of most of the other measurements: they are either much larger or smaller. Outliers can strongly affect certain measures of central tendencies, like the mean.

The ***five-number summary*** is used to describe a set of data using the minimum, quartile 1, median, quartile 3 and the maximum from the data set.

The ***boxplot*** is an easy to read graph of a five-number summary.

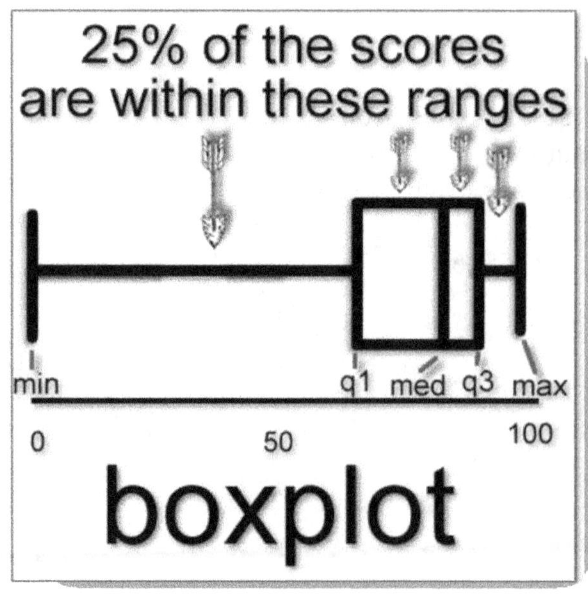

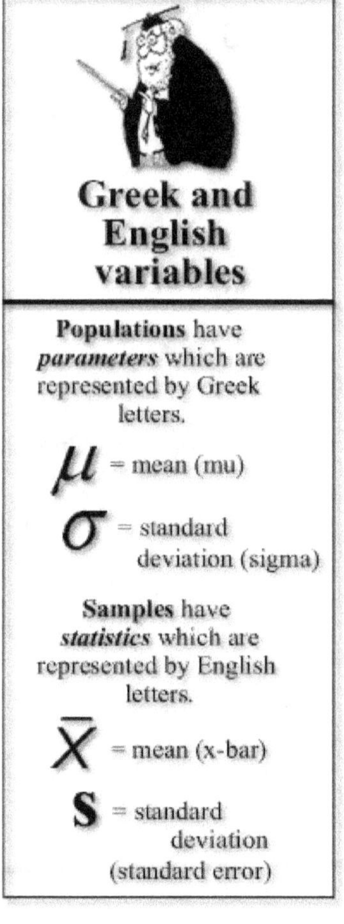

Guided homework Chapter 2: Measures of Central Tendency

The following are the ages of tenured college professors:

57	61	57
57	58	57
61	54	68
51	49	64
50	48	65
52	56	46
54	49	51
47	55	54
42	51	56
55	51	54
51	60	61
43	55	61
52	69	64
46	54	

1. What is the median?

2. What is the mean?

3. What is the mode?

4. What is the variance?

5. What is the standard deviation?

6. What is the five-number summary?

7. Make a boxplot of the observations

Chapter 2 Quiz

The following are test scores from two of Professor Olsen's math class:

0, 0, 18, 22, 26, 41, 42, 52, 53, 59, 60, 60, 61, 64, 64, 65, 67, 70, 70, 71, 72, 72, 74, 74, 75, 77, 78, 78, 80, 81, 85, 86, 88, 88, 90, 92, 95, 97, 100 and 100.

1. What is the mean of the listed test scores?
2. What is the median?
3. What is the mode?
4. What is the variance?
5. What is the standard deviation?
6. What is the first quartile?
7. What is the third quartile?
8. What is the five-number summary?
9. Draw a boxplot of the data set

10. The boxplot below represents salaries of 400 randomly selected people. According to the boxplot, how many people earn under $29,000?

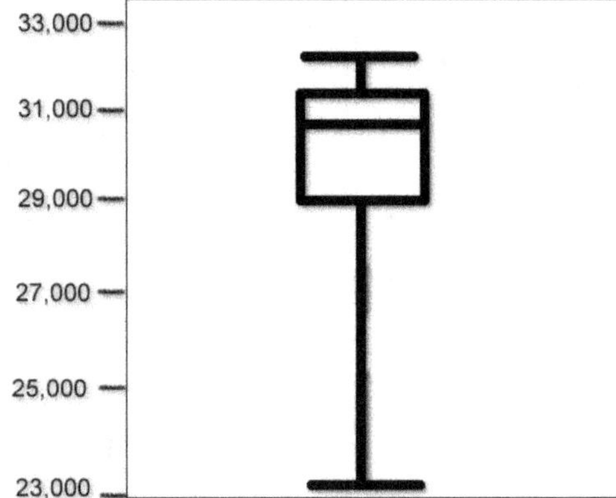

11. The boxplot above represents salaries of 400 randomly selected people. According to the boxplot, around how many people earn between $29,000 and $31,500?

Ch 3: Normal Distributions

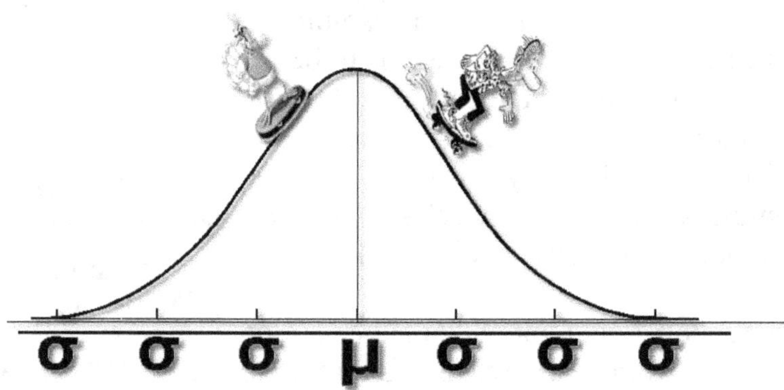

Chapter
3

Normal Distributions

There exists a natural order of measurement to all things. Very few measurements are identical. Some people are bigger than average while others are smaller. Some years see a lot of rain, but some years see very little. When taking measurements of individuals within a set, there will always be a range of values (variance). Very few of the measurements will be equal to each other or exactly equal to the mean. Most of the data will be a little bit larger or smaller than the mean. This spread, or ***distribution***, of the measurements is to be expected, or ***normal***.

Researchers look for three main things in a distribution: overall pattern (symmetrical or skewed?), center and spread. Most of the statistical formulas used to investigate the data are based on the fact that the distributions are normal. If a distribution is **not normal**, then most of the formulas covered in this course would **not be valid**.

In this chapter, we will investigate normal distributions and what we can do with them:

- Random variables
- Density curves
- Normal distributions
- Empirical Rule
- Standardized scores - z scores
- Standard normal distributions
- Normal distribution probability table - z table
- Finding the area under the curve for a specific z score
- Finding the value that will generate a specific proportion (percent)

What's in this chapter

Random Variables

Density Curves

Normal Distributions

Empirical Rule

Standardized Scores

Normal Distribution Table

Probabilty Values

Random variable

In the art of statistics, the concept of a ***random variable*** is an absolute necessity in order for our statistical calculations and conclusions to be valid. Its definition is embedded in its name: ***variable*** means that the value (of the variable) can be ***any*** measurement from the data set, and ***random*** tells us that the measurement has to be selected totally at random without any ***bias*** whatsoever. Bias, in statistics, is cheating. "*Statistics do not lie, but statisticians do.*" – Mark Twain. So, learn your statistics well!

Example of a biased study: A *crooked* statistician got paid a huge sum of money (in cash) to ensure that the results from a survey were favorable to the payer's interests.

The statistician created a questionnaire and performed the poll and claimed that the survey results were: 94% of the 220 people polled stated that they did not want a new tax added on to the price of a pack of cigarettes to help cover health care costs. Here is how the bad statistician biased this survey: he/she waited outside convenience stores and only asked people who had just purchased cigarettes the survey question. He/she knew that cigarette consumers *would not want* to raise the price of their bitter addiction. The moral of this example is "Do not be a crooked statistician!" And don't smoke! Both are bad and dangerous!

Vocabulary

Random variable
A random variable is any measurement from a data set which is selected at random.

Density curve
A density curve is any graph where the total area displayed is considered to be 100%

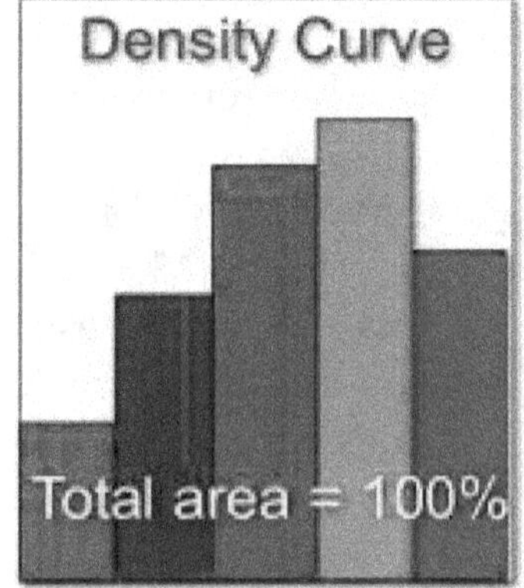

Density curve

A ***density curve*** illustrates the overall pattern of the distribution of a random variable. Density curves are similar to histograms except that the density curve is always on, or above, the horizontal axis and has a total area of 1 (100%). *Not all density curves are normal.*

The area under a density curve represents the **proportion, probability or percent (%)**, of all of the measurements in all categories. The total area under the curve is considered to be 100%. Then each specific category can be compared as a **percentage** to the **whole**. Since the density curve is a distribution of data, it has the same statistical characteristics as lists of numbers, i.e., mean, mode, median, variance, standard deviation and others.

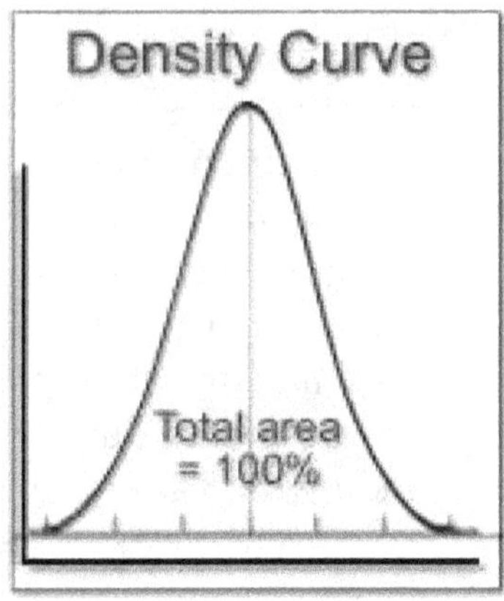

Example 3.1: According to the density curve below, what percent of all 11th graders earned a C in Algebra?

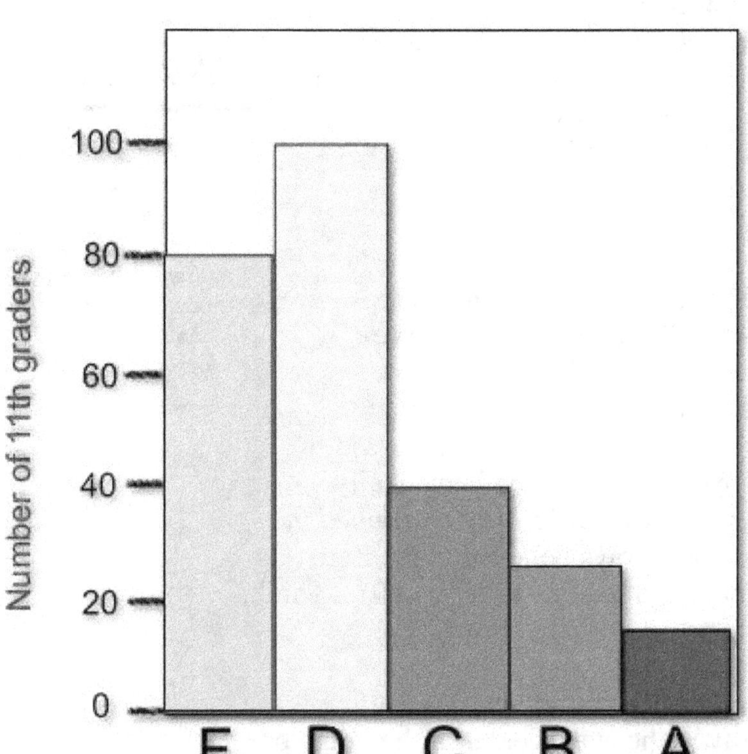

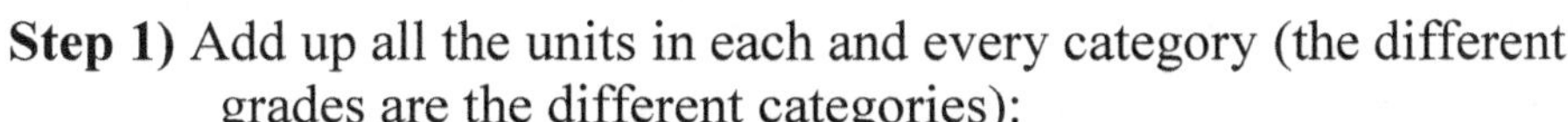

Over 90% of all US 11th graders are not ready for college math.

Step 1) Add up all the units in each and every category (the different grades are the different categories):

80 + 100 + 40 + 30 + 20 +10 = **280**

Step 2) Make a fraction with the *quantity* in the *C category* as the numerator and the sum of all categories as the denominator:

40/280

Step 3) Change the fraction into a percent: 40/280 = **14.3%**

So, 14.3% of this school's 11th graders earned a C in Algebra.

Normal distribution curves

A ***normal distribution curve*** is a **symmetrical** density curve with only one peak or hump. They are also called normal curves, bell curves, and/or ***z curves***. All normal distributions are similar and are described by their means and standard deviations. The value of the mean determines the height of the peak and the value of the standard deviation determines the width.

Vocabulary

Normal distribution
A normal distribution is a symmetrical density curve with only one peak, and the mean, median and mode of the data set are relatively equal to each other.

In a *normal distribution, the mean, median and mode are relatively close to each other*. The **mean is the center of the horizontal axis** and the **standard deviations control the "curving"** of the curve.

A normal distribution curve can be thought of as a symmetrical histogram, but with an infinite number of categories all with the width of a line. The width of a line is zero, bye-the-bye, so mathematically speaking, statistics is the only math that states that ***infinity multiplied by zero is one*** ***(∞ x 0 = 1)***. Freaky! (And you always believed in the Zero Product Rule (that the product of 0 and any number was always 0)... Think again... and keep thinking... mgz out.)

A normal distribution is one that is *relatively* symmetrical. A test for symmetry is to place a mirror vertically in the exact center of the graph of the distribution. The reflected half of the graph should be close to the exact opposite of the half of the graph that is being reflected.

*Note: A normal distribution can be written like this: ***N(μ,σ).*** The capital **N** identifies the distribution as **normal,** ***μ*** is the distribution's **mean** and ***σ*** is the **standard deviation.**

**Note: Sometimes outliers are removed from data sets in order to make the data fit a normal distribution. (Again, it is up to the researcher.)

Here are a few examples of normal distribution curves:

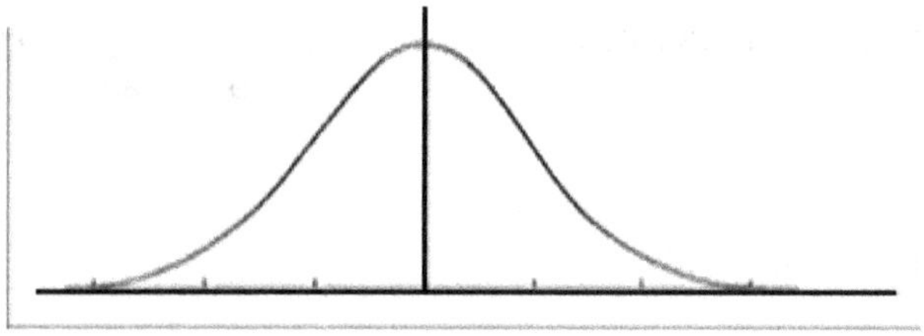

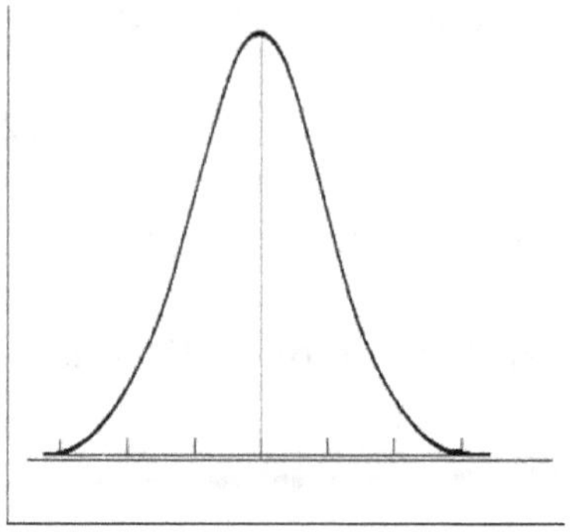

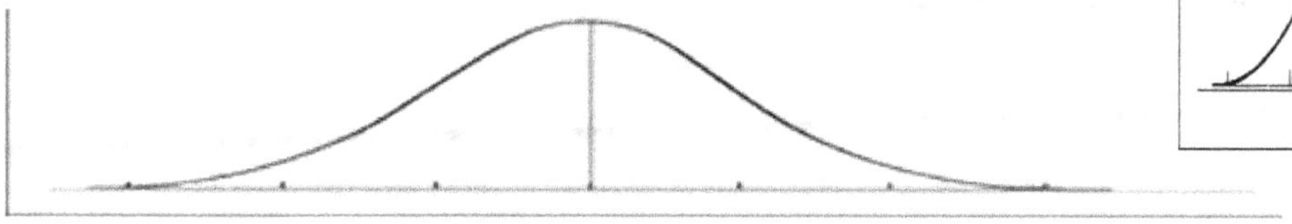

Kurtosis is a term that describes the peak of a distribution. A tall and skinny curve with one or more sharp peaks may be referred to as ***leptokurtotic***, and a wide and low curve with a plateau with no peaks as ***platykurtotic***.

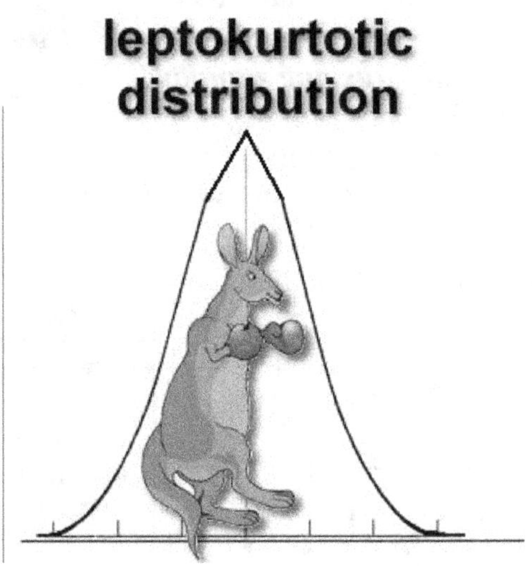

Remember always that not all distributions are normal. When a distribution is lopsided, we call it ***skewed***. Distributions can be skewed to the **right**, or **positively** skewed, or skewed to the *left*, or *negatively* skewed.

A good way to remember the direction of the skew is if the graph was a silhouette of a rat, then the direction of the skew would be the rat's tail. Rats! Ewwwwwwww!

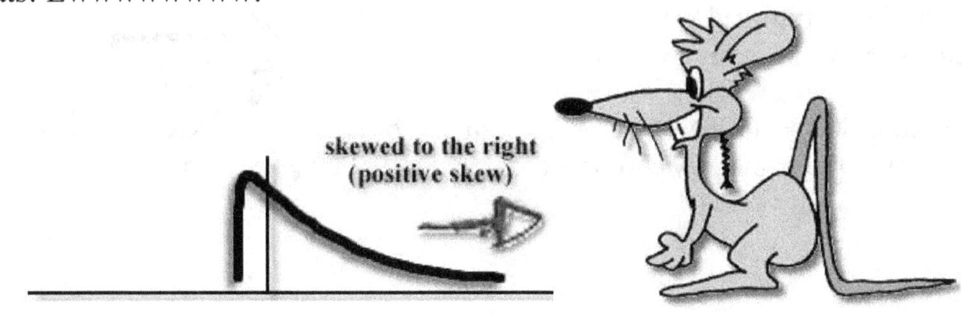

Vocabulary

Skewed distribution
A skewed distribution is not symmetrical! Skew can be described as a left skew (negative skew) or a right skew (positive skew.)

Kurtosis
The kurtosis of a distribution measures the "peakedness" of the distribution. Distributions with tall and thin peaks are **leptokurtotic** and those with short and flat peaks are **platykurtotic**.

Empirical Rule

The ***Empirical Rule*** (also known as the **68%-95%-99.7% Rule**) is a fairly accurate rule-of-thumb estimator for data in a normal distribution. This rule states that ***68%*** of all of the measurements in a normal data set will fall between the range values of the ***mean minus one standard deviation*** (on the low end of the range) and the ***mean plus one standard deviation*** (on the high end), ***95%*** of the data will fall between the mean ± two standard deviations, and ***99.7%*** of the data will fall within the mean ± three standard deviations.

*Note: Remember that half of the measurements will be less than the mean and half will be greater than the mean.

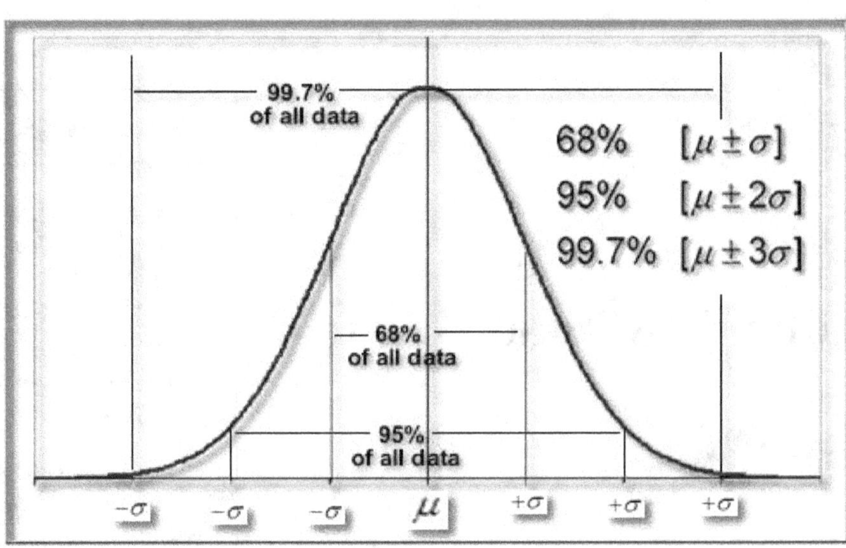

Vocabulary

Range

In mathematics, a range is the difference between the largest measurement and the smallest measurement of a frequency distribution.

To find the range of a set of data, subtract the smallest measurement from the greatest measurement.

A range can aslo be called an **interval**.

Ranges are mathematically written with brackets. Example: if the lowest measurement in a set of data was 12 and the highest was 68, then the range, or interval, would be **[12, 68]**.

Example 3.2: California SAT math test scores fit a normal distribution curve. If the mean score was 500 with a standard deviation of 100, what range of values would 95% of the test scores fall between?

Step 1) Find the mean and standard deviation: **$\mu = 500, \sigma = 100$.**

Step 2) Decide which part of the empirical rule applies: **95%** of the data indicates the **mean ± two standard deviations**.

Step 3) Do the math: $\mu \pm 2\sigma = 500 + 2(100) = 700$ is the high end of the range, $500 - 2(100) = 300$ is the low end of the range. So, 95% of

the SAT math test scores should be between 300 and 700 which can be written in interval form like this: **[300, 700]**.

Using the 68%, 95% and 99.7% percentages from the Empirical Rule, we can calculate other often-used percentages. Here are a few of them:

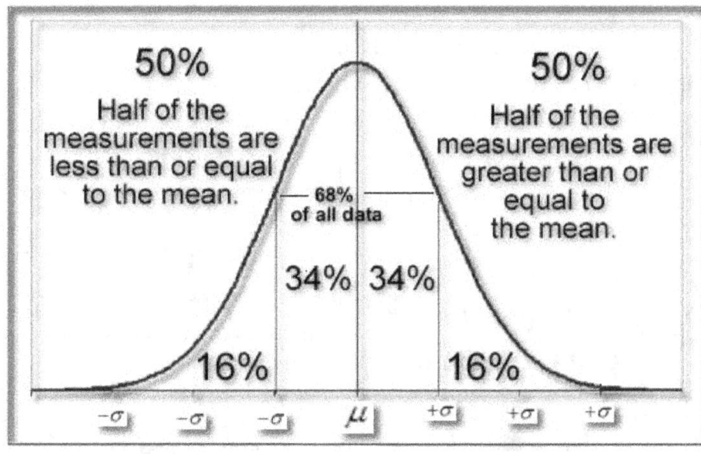

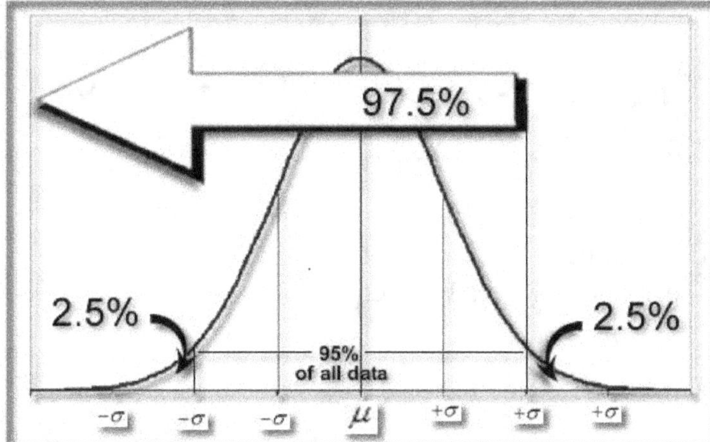

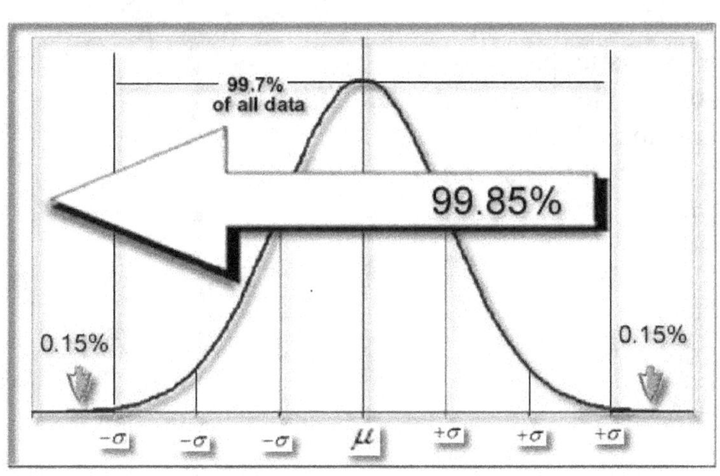

Example 3.3: California SAT test scores fit a normal distribution curve. If the mean score was 500 with a standard deviation of 100, the highest 16% of the test scores would be equal to, or greater than, what score?

Step 1) Find the mean and standard deviation: **μ = 500, σ = 100.**

Step 2) Which part of the empirical rule applies? 16% = 100% - 84%... goes to … 84% = 50% + 34%... and 34% is half of **68%**, which indicates **one standard deviation** greater than the mean.

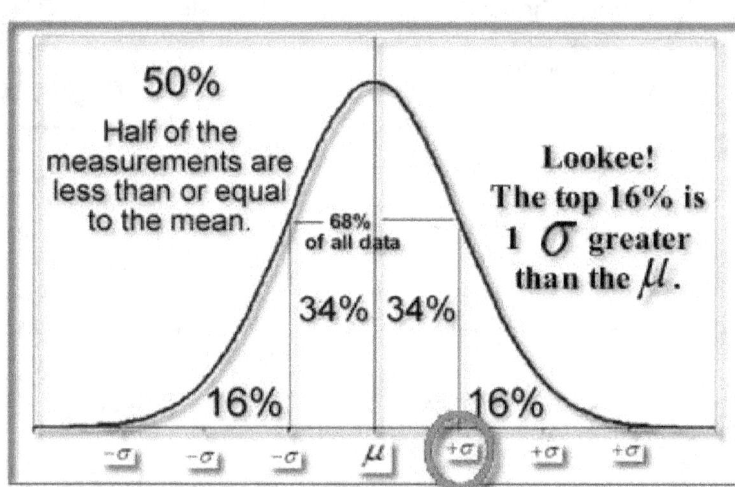

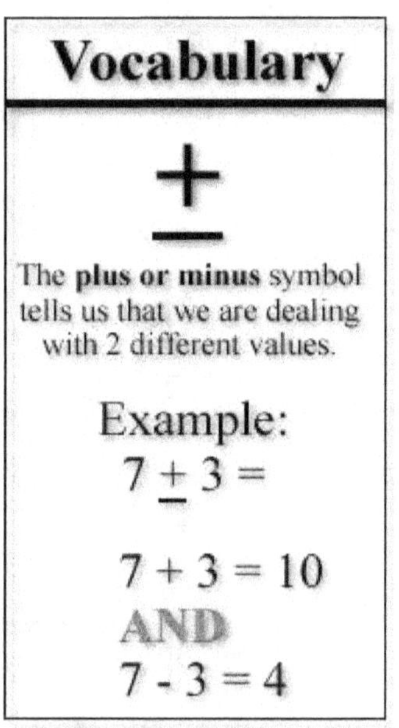

Step 3) Do the math: mean + one standard deviation... **$\mu + \sigma$ = 500 + 100 = 600.** So, 16% of the test scores were equal to, or greater than, **600.**

Standardized scores (z scores)

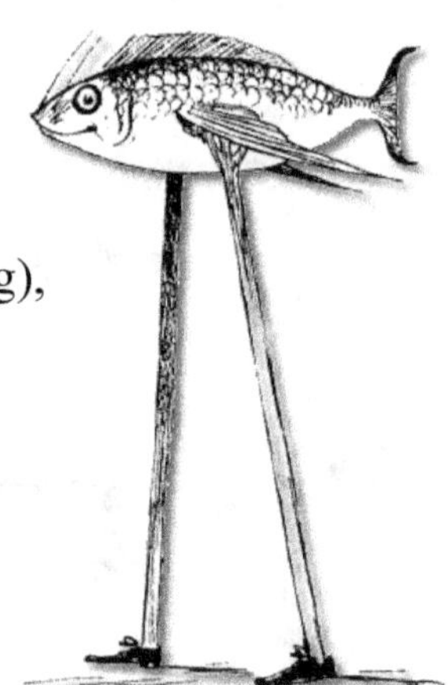

The ***standardized score***, or ***z score***, is a test statistic that is used to compare individual measurements, or groups of measurements (sampling), to the each other and/or to the population mean by telling us *how many standard deviations away that specific measurement, or group measurement, is from the population mean.*

Below is the formula that tells us how many standard deviations away that the individual measurement being investigated (random variable) is away from the mean. (*Note: the formula below is only used when the sample size is 1.)

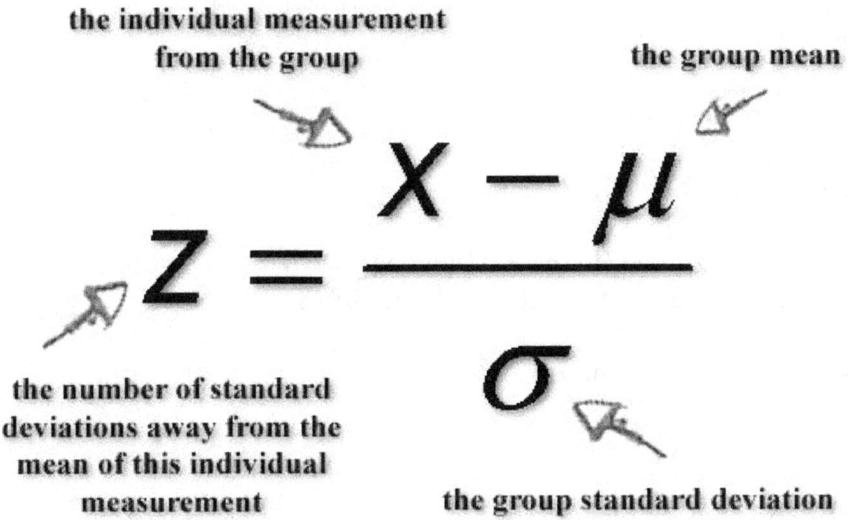

Online video example 3.4: From a normal distribution of test scores with a mean of 85 and a standard deviation of 12, ***N(85, 12)***, what would the standardized score be of a test score of 69?

Step1) Break out the z score formula: $\mathbf{z = (x - \mu)/\sigma}$

Step 2) Substitute the known values: $\mathbf{z = (69 - 85)/12}$

Step 3) Do the math: $z = (69 - 85)/12 = -16/12 = \mathbf{-1.33}$

So, a test score of 69 is 1.33 standard deviations ***less than*** the mean test score. Remember that a ***negative z score*** indicates the number of standard deviations the individual measurement ***is less than the group mean***, and a ***positive z score*** indicates how ***many standard deviations above the mean***.

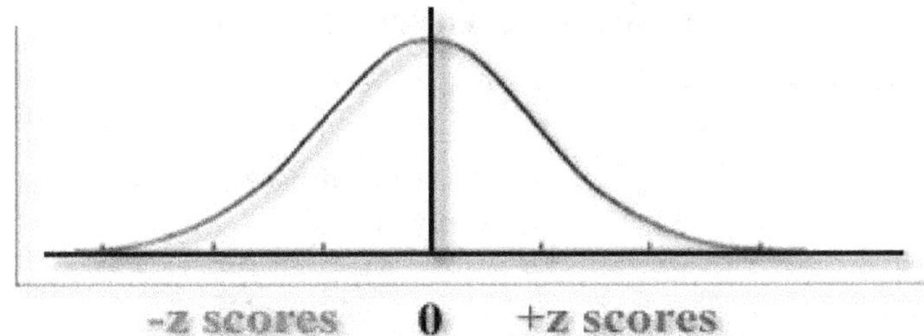

The mean = 0 in a standardized set of data.

Example 3.5: From a normal distribution of test scores with a mean of 85 and a standard deviation of 12, ***N(85, 12)***, what would the probability that a random test score would be ***<u>less than</u>*** 69? This is written like this: ***P(x < 69) =?***

Step1) Break out the z score formula: $\mathbf{z = (x - \mu)/\sigma}$

Step 2) Substitute the known values: $\mathbf{z = (69 - 85)/12}$

Step 3) Do the math: $z = (69 - 85)/12 = -16/12 = \mathbf{-1.33}$

Step 4) Look up the z score of -1.33 in the z table (in the next few pages we will demonstrate looking up p-values from the z table): **.0918**

So, there is a **9.18% probability** that a test score chosen at random will be less than 69.

Standard normal distribution

A ***standard normal distribution*** is a normal distribution with the **mean of 0** and a **standard deviation of 1: N(0, 1)**. Any normal distribution can be converted into a standard distribution by using the standardized formula and converting each and every individual measurement into a standardized score. And we need to remember that the ***total area under this curve is 1***.

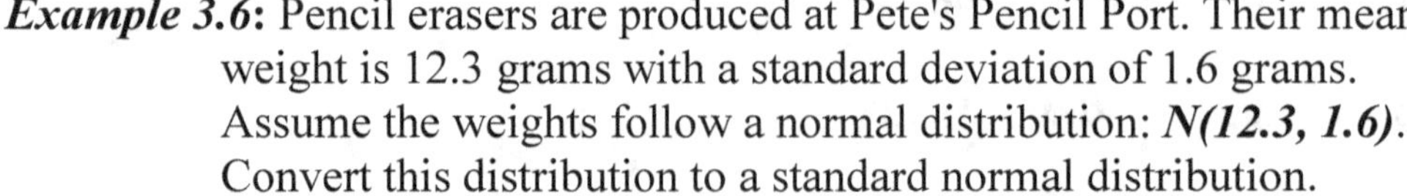
Example 3.6: Pencil erasers are produced at Pete's Pencil Port. Their mean weight is 12.3 grams with a standard deviation of 1.6 grams. Assume the weights follow a normal distribution: ***N(12.3, 1.6)***. Convert this distribution to a standard normal distribution.

Step 1) Easy cheesy: N(12.3, 1.6) is standardized to N(0, 1).

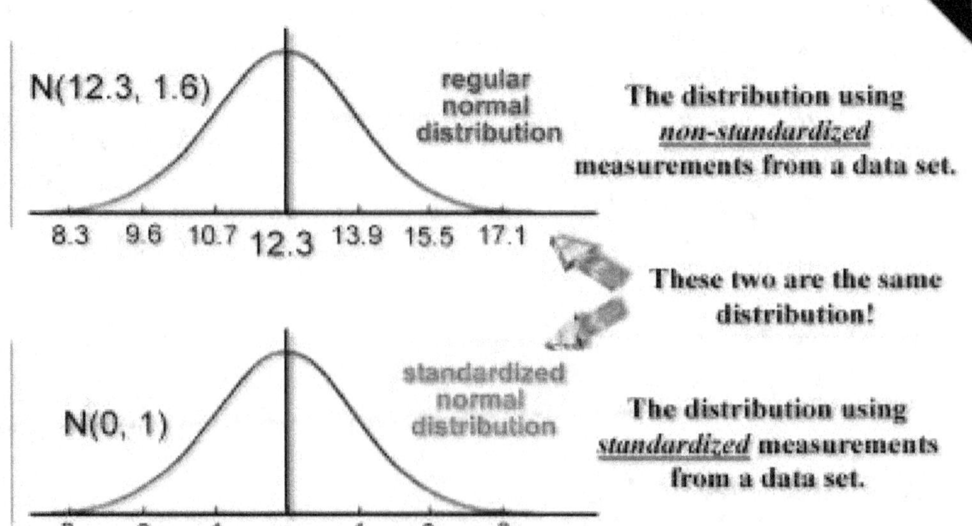

Once an entire set of measurements has been standardized, we can use the calculated z scores to look up their corresponding ***cumulative*** areas in a ***Standard Normal Probabilities*** table, or ***z*** table. The area generated by a specific z score can be found using a *normal distribution table*, or with software programs like Microsoft Excel.

The area under the distribution curve is a proportion of the entire area under the curve. Remember again that the area under the normal distribution curve is always 1. ***This proportional area, or percentage of the whole, is the same as the probability of that random variable being randomly selected.***

The area values, or ***probability values***, from the z table in the textbook gives only the ***cumulative proportions*** (areas or probabilities) for all z score values that are ***less than*** the specific z score. The example below illustrates the p-value from the z table of a z score of -1.56. ***(z < -1.56)***

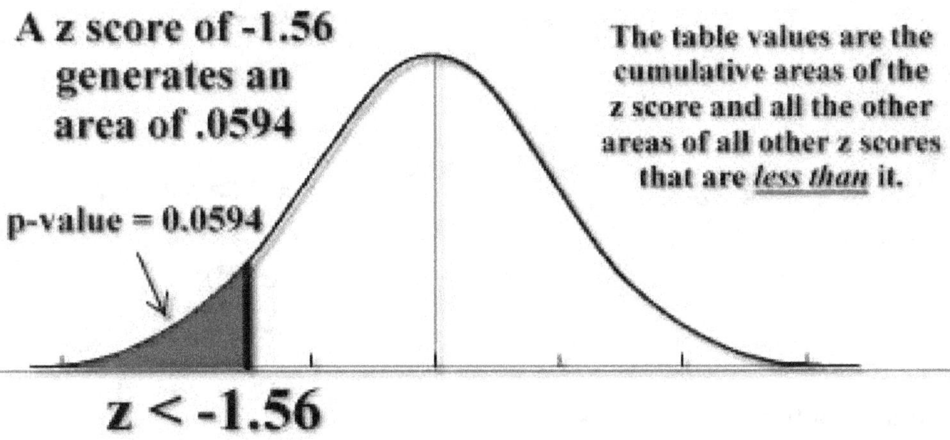

In statistics, the inequality symbols of greater than or equal to (≥) and less than or equal to (≤) are considered to be the same as the greater than (>) and less than (<) symbols, respectively. The reason is because that a specific z score cuts out a corresponding area from the z curve, and that the last vertical "line" has a width of zero. Therefore there is no difference between the ≥ and >, and likewise with the ≤ and <. This is proven with calculus, but that is a different course. Aren't you glad?

If you need the area (proportion or probability) of ***greater than*** a z score, find the corresponding table value from the specific z score, then ***subtract that table value from 1.*** <u>**(*THIS IS A COMMON STUDENT ERROR!)**</u>

Since the entire area is equal to 1, then the following must be true:

Normal Distribution Table

The area under the normal distribution curve represents the probability of the z score.

The Standard Normal Probabilities table values are the ***cumulative*** values, or areas, of the z scores.

Example: the table value for a z score of 1.16 is .8770. So, the probability (area) of the random variable being ***equal to or less than*** 1.16 standard deviations above the group's mean is 87.70%

Cumulative area

The Standard Normal Probabilities table values are the ***cumulative*** values, or areas, of the z scores.

Cumulative area is the sum of all areas, beginning from a z score of - ∞ all the way up to the z score. In other words, the cumulative area of a z score is the proportion of the area under the entire z curve that is on the left of that z score.

Greater than area

$$P(z < x) + P(z > x) = 1$$

and

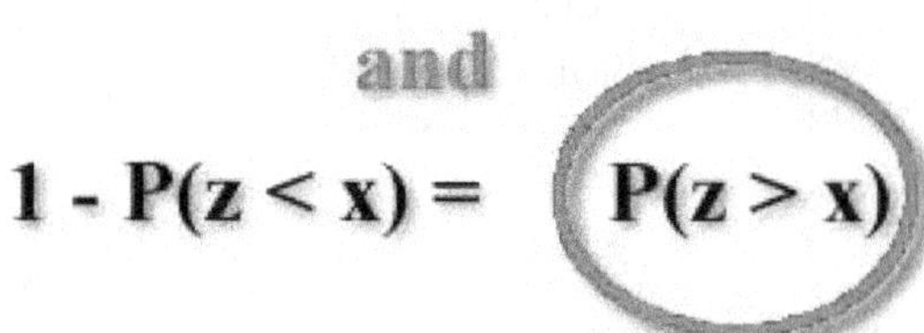

The example below illustrates the p-value from the z table of greater than a z score of -1.56. **(z > -1.56)**

Use the greater than rule!

1 – P(z < -1.56) = P(z > -1.56)
1 – (.0594) = **0.9406 = P(z > -1.56)**

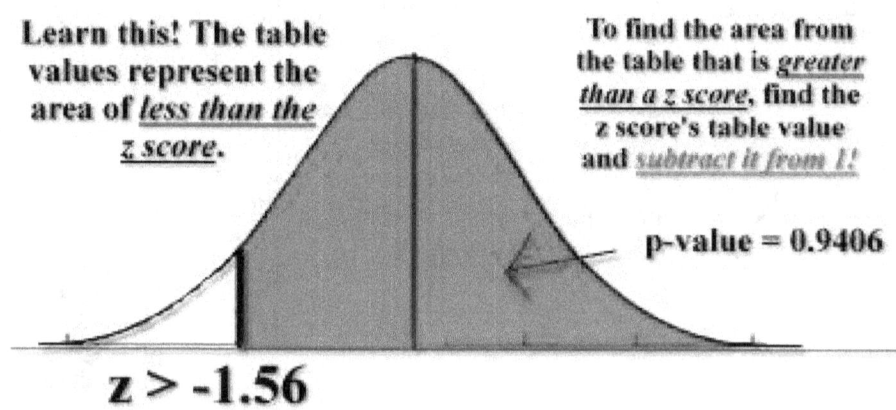

Example 3.7: From a normal distribution of test scores N(85, 12), what would the probability that a random test score would be greater than 69?
(This is written like this: ***P(x > 69)?***)

Example

Step1) Break out the z score formula: **z = (x - μ)/σ**

Step 2) Do the math: z = (69 - 85)/12 = -16/12 = **-1.33**

Step 3) Look up the z score of -1.33 in the z table: **.0918**

Step 4) Subtract .0918 from 1 (because it is a ***greater than***!)
1 - .0918 = .9082. So, there is a **90.82%** chance that a test score selected at random will be higher than 69.

probability value (p-value)

The probability value (p-value) of a specific z score is the area under the standardized normal distribution curve.

Normal Distribution Table

WARNING!
A common student error is to not subtract a table value from 1 when dealing with a GREATER THAN probability!

A correct example:
P(z > -0.63) = ?
(This translates into "What is the area under the graph (probability) ***greater than*** 0.63 standard deviations less than the group's mean?")

Answer: **P(z > -0.63) = ?**
Since the table values only show the cumulative areas (***less than*** a z score) and the total area under the curve is 1, then this must be true...

P(z > -0.63) = 1 - P(z < -0.63)
P(z > -0.63) = 1 - .2643
P(z > -0.63) = .7357

z table

The Standard Normal Probabilities table, or *z* ***table***, is a table of pre-calculated cumulative areas under the z curve (graphically to the left of any specific z score). Please use the z table in the back of this textbook.

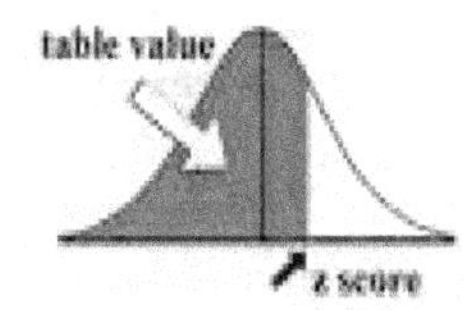

z score
(positive)

Standard Normal Probabilities

	0	0.01	0.02	0.03	0.04	0.05	0.06	0.07	0.08	0.09
0	.5000	.5040	.5080	.5120	.5160	.5199	.5239	.5279	.5319	.5359
0.1	.5398	.5438	.5478	.5517	.5557	.5596	.5636	.5675	.5714	.5753
0.2	.5793	.5832	.5871	.5910	.5948	.5987	.6026	.6064	.6103	.6141
0.3	.6179	.6217	.6255	.6293	.6331	.6368	.6406	.6443	.6480	.6517
0.4	.6554	.6591	.6628	.6664	.6700	.6736	.6772	.6808	.6844	.6879
0.5	.6915	.6950	.6985	.7019	.7054	.7088	.7123	.7157	.7190	.7224
0.6	.7257	.7291	.7324	.7357	.7389	.7422	.7454	.7486	.7517	.7549
0.7	.7580	.7611	.7642	.7673	.7704	.7734	.7764	.7794	.7823	.7852
0.8	.7881	.7910	.7939	.7967	.7995	.8023	.8051	.8078	.8106	.8133
0.9	.8159	.8186	.8212	.8238	.8264	.8289	.8315	.8340	.8365	.8389
1	.8413	.8438	.8461	.8485	.8508	.8531	.8554	.8577	.8599	.8621
1.1	.8643	.8665	.8686	.8708	.8729	.8749	.8770	.8790	.8810	.8830
1.2	.8849	.8869	.8888	.8907	.8925	.8944	.8962	.8980	.8997	.9015
1.3	.9032	.9049	.9066	.9082	.9099	.9115	.9131	.9147	.9162	.9177
1.4	.9192	.9207	.9222	.9236	.9251	.9265	.9279	.9292	.9306	.9319
1.5	.9332	.9345	.9357	.9370	.9382	.9394	.9406	.9418	.9429	.9441
1.6	.9452	.9463	.9474	.9484	.9495	.9505	.9515	.9525	.9535	.9545
1.7	.9554	.9564	.9573	.9582	.9591	.9599	.9608	.9616	.9625	.9633
1.8	.9641	.9649	.9656	.9664	.9671	.9678	.9686	.9693	.9699	.9706
1.9	.9976	.9975	.9726	.9732	.9738	.9744	.9750	.9756	.9761	.9767
2	.9772	.9778	.9783	.9788	.9793	.9798	.9803	.9808	.9812	.9817
2.1	.9821	.9826	.9830	.9834	.9838	.9842	.9846	.9850	.9854	.9857
2.2	.9861	.9864	.9868	.9871	.9875	.9878	.9881	.9884	.9887	.9890
2.3	.9893	.9896	.9898	.9901	.9904	.9906	.9909	.9911	.9913	.9916
2.4	.9918	.9920	.9922	.9925	.9927	.9929	.9931	.9932	.9934	.9936
2.5	.9938	.9940	.9941	.9943	.9945	.9946	.9948	.9949	.9951	.9952
2.6	.9953	.9955	.9956	.9957	.9959	.9960	.9961	.9962	.9963	.9964
2.7	.9965	.9966	.9967	.9968	.9969	.9970	.9971	.9972	.9973	.9974
2.8	.9974	.9975	.9976	.9977	.9977	.9978	.9979	.9979	.9980	.9981
2.9	.9981	.9982	.9982	.9983	.9984	.9984	.9985	.9985	.9986	.9986
3	.9987	.9987	.9987	.9988	.9988	.9989	.9989	.9989	.9990	.9990
3.1	.9990	.9991	.9991	.9991	.9992	.9992	.9992	.9992	.9993	.9993
3.2	.9993	.9993	.9994	.9994	.9994	.9994	.9994	.9995	.9995	.9995
3.3	.9995	.9995	.9995	.9996	.9996	.9996	.9996	.9996	.9996	.9997
3.4	.9997	.9997	.9997	.9997	.9997	.9997	.9997	.9997	.9997	.9998

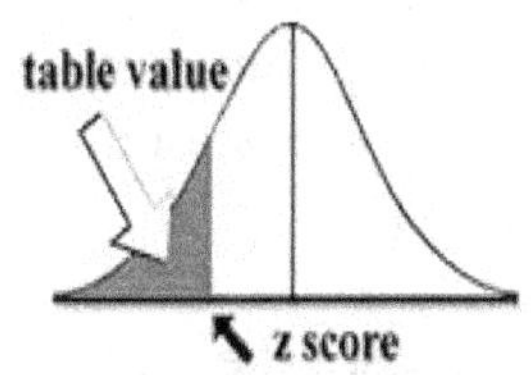

z score
(negative)

Standard Normal Probabilities

	0	0.01	0.02	0.03	0.04	0.05	0.06	0.07	0.08	0.09
0	.5000	.4960	.4920	.4880	.4840	.4801	.4761	.4721	.4681	.4641
-0.1	.4602	.4562	.4522	.4483	.4443	.4404	.4364	.4325	.4286	.4247
-0.2	.4207	.4168	.4129	.4090	.4052	.4013	.3974	.3936	.3897	.3859
-0.3	.3821	.3783	.3745	.3707	.3669	.3632	.3594	.3557	.3520	.3483
-0.4	.3446	.3409	.3372	.3336	.3300	.3264	.3228	.3192	.3156	.3121
-0.5	.3085	.3050	.3015	.2981	.2946	.2912	.2877	.2843	.2810	.2776
-0.6	.2743	.2709	.2676	.2643	.2611	.2578	.2546	.2514	.2483	.2451
-0.7	.2420	.2389	.2358	.2327	.2296	.2266	.2236	.2206	.2177	.2148
-0.8	.2119	.2090	.2061	.2033	.2005	.1977	.1949	.1922	.1894	.1867
-0.9	.1841	.1814	.1788	.1762	.1736	.1711	.1685	.1660	.1635	.1611
-1	.1587	.1562	.1539	.1515	.1492	.1469	.1446	.1423	.1401	.1379
-1.1	.1357	.1335	.1314	.1292	.1271	.1251	.1230	.1210	.1190	.1170
-1.2	.1151	.1131	.1112	.1093	.1075	.1056	.1038	.1020	.1003	.0985
-1.3	.0968	.0951	.0934	.0918	.0901	.0885	.0869	.0853	.0838	.0823
-1.4	.0808	.0793	.0778	.0764	.0749	.0735	.0721	.0708	.0694	.0681
-1.5	.0668	.0655	.0643	.0630	.0618	.0606	.0594	.0582	.0571	.0559
-1.6	.0548	.0537	.0526	.0516	.0505	.0495	.0485	.0475	.0465	.0455
-1.7	.0446	.0436	.0427	.0418	.0409	.0401	.0392	.0384	.0375	.0367
-1.8	.0359	.0351	.0344	.0336	.0329	.0322	.0314	.0307	.0301	.0294
-1.9	.0287	.0281	.0274	.0268	.0262	.0256	.0250	.0244	.0239	.0233
-2	.0228	.0222	.0217	.0212	.0207	.0202	.0197	.0192	.0188	.0183
-2.1	.0179	.0174	.0170	.0166	.0162	.0158	.0154	.0150	.0146	.0143
-2.2	.0139	.0136	.0132	.0129	.0125	.0122	.0119	.0116	.0113	.0110
-2.3	.0107	.0104	.0102	.0099	.0096	.0094	.0091	.0089	.0087	.0084
-2.4	.0082	.0080	.0078	.0075	.0073	.0071	.0069	.0068	.0066	.0064
-2.5	.0062	.0060	.0059	.0057	.0055	.0054	.0052	.0051	.0049	.0048
-2.6	.0047	.0045	.0044	.0043	.0041	.0040	.0039	.0038	.0037	.0036
-2.7	.0035	.0034	.0033	.0032	.0031	.0030	.0029	.0028	.0027	.0026
-2.8	.0026	.0025	.0024	.0023	.0023	.0022	.0021	.0021	.0020	.0019
-2.9	.0019	.0018	.0018	.0017	.0016	.0016	.0015	.0015	.0014	.0014
-3	.0013	.0013	.0013	.0012	.0012	.0011	.0011	.0011	.0010	.0010
-3.1	.0010	.0009	.0009	.0009	.0008	.0008	.0008	.0008	.0007	.0007
-3.2	.0007	.0007	.0006	.0006	.0006	.0006	.0006	.0005	.0005	.0005
-3.3	.0005	.0005	.0005	.0004	.0004	.0004	.0004	.0004	.0004	.0003
-3.4	.0003	.0003	.0003	.0003	.0003	.0003	.0003	.0003	.0003	.0002

*Note: z scores are rounded to the nearest hundredth (two decimal place values) in a typical standard normal distribution table. The ones' digit and the tenths' digit are in the columns on the left side of the z table while the hundredths' digits run horizontally on top of the table.

Normal Distribution Table

****WARNING!****
A common student error is to not subtract a table value from 1 when dealing with a GREATER THAN probability!

<u>A correct example:</u>
P(z > -0.63) = ?
(This translates into "What is the area under the graph (probability) ***greater than*** 0.63 standard deviations less than the group's mean?")

Answer: **P(z > -0.63) = ?**
Since the table values only show the cumulative areas (***<u>less than</u>*** a z score) and the total area under the curve is 1, then this must be true...

P(z > -0.63) = 1 - P(z < -0.63)
P(z > -0.63) = 1 - .2643
P(z > -0.63) = .7357

To find a table value from a z score with two decimals, find the leading digits in the z score column, run your finger across that row until it intersects with the corresponding hundredths column. That number is the correct p-value.

Here is an example of the table p-value of a z score of -1.28. 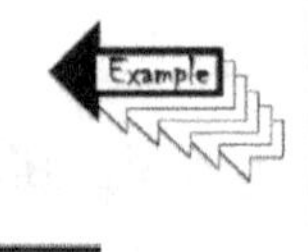

z score
(negative)

z scor

Standard Normal Probabilities

	0	0.01	0.02	0.03	0.04	0.05	0.06	0.07	0.08	0.09
0	.5000	.4960	.4920	.4880	.4840	.4801	.4761	.4721	.4681	.4641
-0.1	.4602	.4562	.4522	.4483	.4443	.4404	.4364	.4325	.4286	.4247
-0.2	.4207	.4168	.4129	.4090	.4052	.4013	.3974	.3936	.3897	.3859
-0.3	.3821	.3783	.3745	.3707	.3669	.3632	.3594	.3557	.3520	.3483
-0.4	.3446	.3409	.3372	.3336	.3300	.3264	.3228	.3192	.3156	.3121
-0.5	.3085	.3050	.3015	.2981	.2946	.2912	.2877	.2843	.2810	.2776
-0.6	.2743	.2709	.2676	.2643	.2611	.2578	.2546	.2514	.2483	.2451
-0.7	.2420	.2389	.2358	.2327	.2296	.2266	.2236	.2206	.2177	.2148
-0.8	.2119	.2090	.2061	.2033	.2005	.1977	.1949	.1922	.1894	.1867
-0.9	.1841	.1814	.1788	.1762	.1736	.1711	.1685	.1660	.1635	.1611
-1	.1587	.1562	.1539	.1515	.1492	.1469	.1446	.1423	.1401	.1379
-1.1	.1357	.1335	.1314	.1292	.1271	.1251	.1230	.1210	.1190	.1170
-1.2	.1151	.1131	.1112	.1093	.1075	.1056	.1038	.1020	.1003	.0985
-1.3	.0968	.0951	.0934	.0918	.0901	.0885	.0869	.0853	.0838	.0823
-1.4	.0808	.0793	.0778	.0764	.0749	.0735	.0721	.0708	.0694	.0681

So the probability value (p-value) of a z score of -1.28 is **0.1003**.

There are three different types of these problems. The following are examples of each type.

Online video example 3.8: (Type 1 - ***<u>less than</u>***)

What is the area generated by a z score of -1.33?

Step1) Rewrite it as P(z < -1.33)?

Step 2) Look up the table value of -1.33: **.0918**

Step 3) You are done because it is a ***less than*** problem.

Online video example 3.9: (Type 2 – ***greater than***)

What is P(z > -2.17)?

Step1) Look up the table value of -2.17: **.0150**

Step 2) Subtract that table value from 1: **1 - . 0150 = .9850**

Step 3) You are done because it is a ***greater than*** problem.

Online video example 3.10: (Type 3 – ***between***)

What is P(-1.35 < z < 0.62)?

Step1) Look up the table value of **larger** z score **first**:
z of 0.62 gives us an area (p-value) of **.7324**

Step 2) Look up the table value of *smaller* z score *second*:
z of -1.35 gives us an area of **.0885**

Step 3) Subtract the smaller area from the larger area:
.7324 - .0885 = **.6439** (Done!)

Finding an individual measurement given a proportion or percent first

Sometimes you will have to find a specific measurement from a data set that is a cut-off value representing a specific percent, or a proportion, of a normal distribution curve. We do this by using the z formula and the z table.

Online video example 3.11: Farmer Ted grows several different varieties of corn. He is testing a new hybrid seed and wants to gather some data about the results. After the harvest, Ted measures the new corn stalks. The distribution of the height of corn stalks is normal with a mean of 2.7 meters and a standard deviation of 0.4 meters. Ten percent of the shortest stalks will be equal to or less than what height?

Step 1) Look in the z table for an area (*not a z score but a p-value!*) that is close to 10%, or 0.1000. *The closest in the z table is an area (p-value) of 0.1003.*

Step 2) Find the z score that generates that area: an area of 0.1003 is generated by a z score of -1.28.

z score
(negative)

Standard Normal Probabilities

z sco

	0	0.01	0.02	0.03	0.04	0.05	0.06	0.07	0.08	0.09
0	.5000	.4960	.4920	.4880	.4840	.4801	.4761	.4721	.4681	.4641
-0.1	.4602	.4562	.4522	.4483	.4443	.4404	.4364	.4325	.4286	.4247
-0.2	.4207	.4168	.4129	.4090	.4052	.4013	.3974	.3936	.3897	.3859
-0.3	.3821	.3783	.3745	.3707	.3669	.3632	.3594	.3557	.3520	.3483
-0.4	.3446	.3409	.3372	.3336	.3300	.3264	.3228	.3192	.3156	.3121
-0.5	.3085	.3050	.3015	.2981	.2946	.2912	.2877	.2843	.2810	.2776
-0.6	.2743	.2709	.2676	.2643	.2611	.2578	.2546	.2514	.2483	.2451
-0.7	.2420	.2389	.2358	.2327	.2296	.2266	.2236	.2206	.2177	.2148
-0.8	.2119	.2090	.2061	.2033	.2005	.1977	.1949	.1922	.1894	.1867
-0.9	.1841	.1814	.1788	.1762	.1736	.1711	.1685	.1660	.1635	.1611
-1	.1587	.1562	.1539	.1515	.1492	.1469	.1446	.1423	.1401	.1379
-1.1	.1357	.1335	.1314	.1292	.1271	.1251	.1230	.1210	.1190	.1170
-1.2	.1151	.1131	.1112	.1093	.1075	.1056	.1038	.1020	.1003	.0985
-1.3	.0968	.0951	.0934	.0918	.0901	.0885	.0869	.0853	.0838	.0823
-1.4	.0808	.0793	.0778	.0764	.0749	.0735	.0721	.0708	.0694	.0681

Step 3) Pull out the z score formula:

$$z = \frac{x - \mu}{\sigma}$$

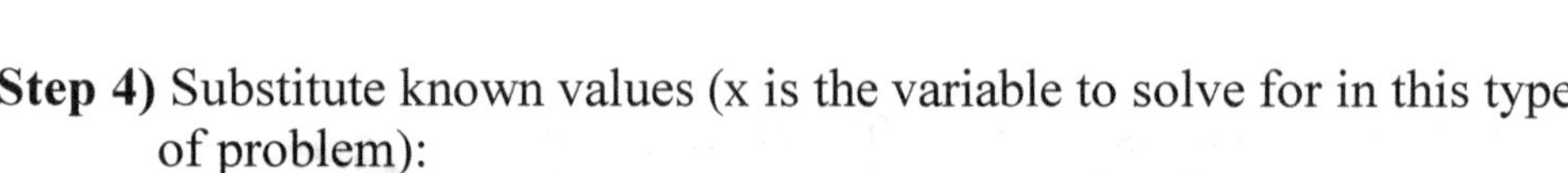

Step 4) Substitute known values (x is the variable to solve for in this type of problem):

$$-1.28 = \frac{x - 2.7}{0.4}$$

Step 5) Cross multiply and divide:

$$-1.28(0.4) = 1(x - 2.7)$$

$$-0.512 = x - 2.7$$

$$2.7 - .512 = x$$

$$2.188 = x$$

The shortest 10% of the stalks are equal to or shorter than **2.188** meters.

Example 3.12: Farmer Ted grows several different varieties of corn. He is testing a new hybrid seed and wants to gather some data about the results. After the harvest, Ted measures the new corn stalks. The distribution of the height of corn stalks is normal with a mean of 2.7 meters and a standard deviation of 0.4 meters. The ***tallest*** 25% of the stalks will be equal to, or greater than, what height?

Step 1) Look in the z table for an area (a table value, not a z score!) that is closest to 75% or 0.7500 - ***because this is a greater than problem***! Remember that the areas from the z table always represent the cumulative z score, (the sum of p-values less than a z score) so we subtract: 100% - 25% = 75%. The z score that generates ***25% greater than*** is the same z score that generates ***75% less than***. The closest in the z table is an area (p-value) of 0.7486.

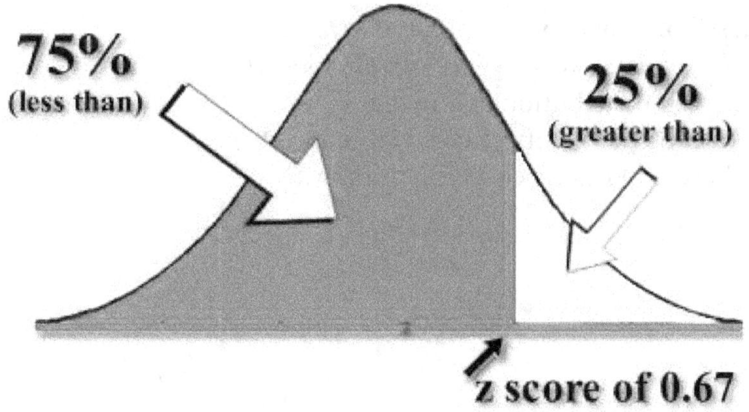

Step 2) Find the z score that generates that area:
z of 0.67 generates an area of **0.7486**

Step 3) Pull out the z score formula: $\mathbf{z = (x - \mu)/\sigma}$

Step 4) Substitute known values (x is the variable to solve for in this type of problem): $z = (x - \mu)/\sigma$ ... $0.67 = (x - 2.7)/0.4$

Step 5) Cross multiply and divide ... $0.67 = (x - 2.7)/0.4$...
$0.67(0.4) = (1)(x - 2.7)$ ...
$0.268 = x - 2.7$ ...
$\mathbf{2.968 = x}$

The tallest 25% stalks are **2.968** meters or taller.

Online video examples 3.13 – 3.18

Cumulative area

The Standard Normal Probabilities table values are the ***cumulative*** values, or areas, of the z scores.

Cumulative area is the sum of all areas, beginning from a z score of $-\infty$ all the way up to the z score. In other words, the cumulative area of a z score is the proportion of the area under the entire z curve that is on the left of that z score.

Chapter 3 Review

A ***random variable*** can be ***any*** measurement chosen randomly from the data set.

A ***density curve*** is used to show the overall pattern of a distribution. The **total area of a density curve is 1**, or 100%. The different categories in a density curve show the percentage, or proportion, of the measurements that fall into that specific category compared to the entire set of measurements.

Chapter Review

A ***normal distribution*** is a symmetrical density curve with only one peak, or hump, with half of the measurements less than the group's mean and half of the measurements greater than the mean.

Kurtosis describes the "peakedness" of a distribution. ***Leptokurtotic*** is a distribution with one or more pointed peaks, while ***platykurtotic*** describes a distribution with a plateau. A ***skewed*** distribution is one that is not symmetric. Skewed distributions can be positively skewed (to the right) or negatively skewed (to the left). These conditions could invalidate a study.

The ***Empirical Rule*** (also known as the **68% - 95% - 99.7% Rule**) is a fairly accurate rule-of-thumb estimator for data in a ***normal*** distribution. This rule states that ***68%*** of all of the measurements in a normal data set will fall between the range values of the ***mean minus one standard deviation*** (on the low end of the range) and the ***mean plus one standard deviation*** (on the high end), ***95%*** of the data will fall between the mean ± two standard deviations, and ***99.7%*** of the data will fall within the mean ± three standard deviations.

In a ***normal*** distribution, the Empirical Rule tells us how many of the measurements fall within specific ranges of values, depending on the mean and the standard deviation.

68% $[\mu \pm \sigma]$

95% $[\mu \pm 2\sigma]$

99.7% $[\mu \pm 3\sigma]$

The ***standardized score***, or ***z score***, is a test statistic that is used to compare individual measurements, or groups of measurements (sampling), to the each other and/or the population mean by telling us how many standard deviations away that specific measurement, or group measurement, is from the population mean.

A ***standard normal distribution,*** or a z distribution**,** is a normal distribution with the mean of 0 and a standard deviation of 1: N(0, 1). Any normal distribution can be converted into a standard distribution by using the standardized formula and converting each and every individual measurement into a standardized score.

The ***Standard Normal Distribution table***, ***Standard Normal Probabilities table,*** or ***z table***, is a table of pre-calculated ***cumulative*** areas (p-values) under the z curve (graphically to the left of any specific z score).

Guided homework Chapter 3: Normal Distributions

The weights (in grams) of Reese's Pieces candies fit a normal distribution curve with N(0.781, 0.026).

1. 25% of the candies will weigh under what weight?

2. What is the probability that an individual piece (of Reese) will weigh under 0.786 grams?

3. What is the probability that an individual piece (of Reese) will weigh over 0.795 grams?

4. The top 10% of the candies will weigh over what weight?

5. What percentage of the pieces will weigh between 0.765 and 0.785 grams?

6. 95% of the candy will be between what two weights?

Chapter 3 Quiz

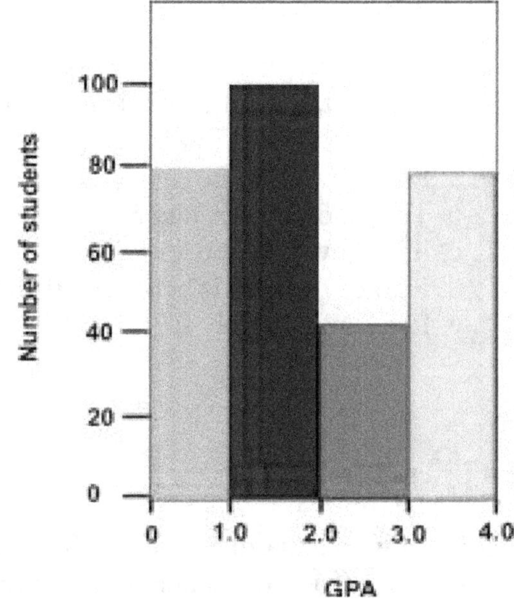

1. According to the density curve to the right, what percentage of students have a GPA between 1.0 and 2.0?

2. What is $P(z < 0.71)$? (Use the z table.)

3. What is $P(x > 2.18)$? (Use the z table.)

4. The Calico Cat Company has a truckload o' kittens with weights that are normally distributed. Their mean weight is 2.40 pounds with a standard deviation of 0.36 pounds.

a. What percent of the kittens will weigh between 2.1 and 2.6 lbs?

b. 68% of the kittens will weigh between what two weights?

c. The heaviest 10% of the kittens will weigh over what weight?

d. 50% of the kittens will weigh over what weight?

e. What is the probability that a randomly chosen kitten will weigh over 3 pounds?

f. The smallest 5% of the kittens will weigh under what weight?

5. If you standardized 100 test scores, what number would the sum of all those z scores be close to?

6. Which statement below is true about the boxplot?

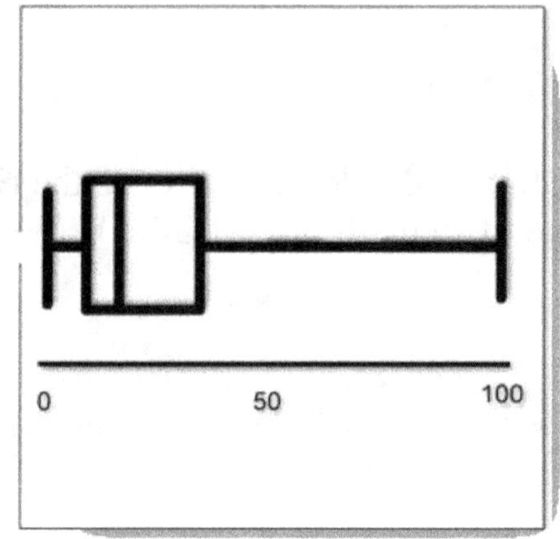

a. It is skewed to the right
b. The median is around 50
c. It is skewed to the left
d. Half of the data is between 50 and 100
e. It is symmetrical

7. True or false?

When we standardize a set of measurements, the mean would be 0 and the standard deviation would be 1.

8. True or false?

$P(x > 1.93) = 1 - P(x < 1.93)$

9. The distribution of weights (in ounces) of "The Big Fudgy" ice cream bar is approximately normal with N(4.27, 0.33).

a. What percent of the ice cream bars weigh less than 3.94 ounces?

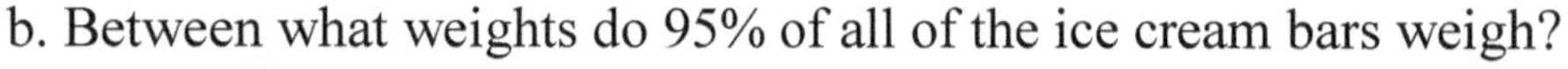

b. Between what weights do 95% of all of the ice cream bars weigh?

c. 2.5% of all of the ice cream bars will weigh under what weight?

10. From a normal distribution of test scores with a mean of 72 and a standard deviation of 12, N(72, 12), what would the standardized test score of 72 be?

11. Tammy and Pammy are identical twin sisters with bright red hair and freckles... many many freckles. Tammy scored 650 on the math section of the SAT test which has the distribution of N(560, 60). Pammy scored a 22 on the ACT test which has the following distribution N(17, 2).

Who had the higher standardized test score?

Ch 4: Correlations and Lines of Regression

Chapter 4

Correlations and Lines of Regression

A fisherman was trying out new bait that supposedly attracts more fish than other baits. He experimented with the amount of bait and kept track of the time it took before he got a nibble. He was very satisfied because it seemed to him that the more bait that he put on the hook, the faster a fish would take a nibble. This ***relationship between the two variables*** (*amount of bait* and the *time* it took a fish to come and nibble) is called a ***correlation***.

A correlation is a statistical tool that measures the *strength* and *direction* of a relationship between two quantitative variables. In statistics we use the correlation to investigate if one variable is influencing, or *controlling*, the other variable.

In this chapter, we will investigate correlations:

- Explanatory and response variables
- Scatterplots
- Best fit line
- Measuring correlation with a formula
- Lines of regression
- Least-squares line of regression
- Residuals

What's in this chapter

Explanatory and response variables

The ***strength*** and ***direction*** of a relationship between two quantitative variables are what statisticians are looking for in a ***correlation***. Is one variable influencing the other? If so, the influencing variable is called the ***explanatory*** variable because it *explains* what is happening to the other variable. The variable that is being influenced is called the ***response*** variable because it is simply *responding* to the explanatory variable. Their names define their roles in a correlation.

Explanatory variables also go by the name of ***independent variables***, factors, predictors, causes, and other names that mean roughly the same as *controller variable*.

Response variables also go by the name of ***dependent variables***, outcomes, effects, criterion variables and other names that mean roughly the same as *controlled variable*.

*Note: it is not always easy to determine which variable is the explanatory and which is the response. Be a careful statistician and **never assume absolute causality**. Remember that ***correlation does not prove causality***!

Vocabulary

Correlation
A correlation is a statistical tool that measures the strength of a relationship between two variables.

Explanatory variable
In a correlation, an explanatory variable is the controlling variable. It "explains" what is happening in a relationship.

Response variable
In a correlation, a response variable is the variable being "controlled" by the explanatory variable.

Example 4.1: In the examples below, which is the explanatory variable and which is the response variable?

Q 1) The temperature of a pot of tea compared to the height of the flame underneath the teapot.

A 1) The height of the flame is the explanatory variable because it explains the temperature of the tea, which is the response variable.

Q 2) The number of hours worked compared to your paycheck.

A 2) The number of hours worked is the explanatory variable because it explains how large (or small) your paycheck is, which is the response variable.

*Note: Researchers must be careful when trying to prove causality through a correlation because there may be a different unseen and untested variable affecting both the explanatory and the response variables. This unseen force is called a "***lurking***" variable because it lurks in the shadows (insert spooky laugh here). A lurking variable can also be referred to a "***confounding***" variable.

Q 3) The number of umbrellas sold compared to how many bottles of flu medicine sold.

A 3) Is the explanatory variable the number of umbrellas sold? If this were true, then the response variable would be the number of bottles of flu medicine sold. You would be stating that umbrella purchases cause people to buy flu medicine. Or is it the other way around: do flu medicine sales cause umbrella sales? Stop! This is crazy talk!

What is truly going on in example 3 is that there is a ***lurking*** variable that is influencing both of the other variables. The lurking variable in this problem is probably the cold weather that winter brings... brrrrrrrr!

Scatterplots

A ***scatterplot*** is a *graph of the relationship between two variables.* Usually the explanatory variable, if one exists, is plotted on the x axis and the response variable is plotted on the y axis.

Scatterplots are graphed using the coordinate system on the Cartesian plane. Remember how to graph coordinates, or points, on a coordinate graph? *The points on a scatterplot are actual (real) recorded data.*

Each point on the scatterplot represents a unique value of the explanatory variable and its relationship with the corresponding value of the response variable.

*Note: If you are not sure which variable is the explanatory variable, any variable can be the x axis or the y axis. Believe it or not, the correlation would be the exact same!

Example 4.2: Migratory sparrows were counted for thirteen years. The percent of birds from the original flock was recorded along with the number of new sparrows that had recently joined the flock. Make a ***scatterplot*** that graphically represents the percent of birds from the original flock and the number of new birds.

percent	new birds
74	5
66	6
81	8
52	11
73	12
62	15
52	16
45	17
62	18
46	18
60	19
46	20
38	20

Step 1) Draw a coordinate graph with x and y axes.

Steps 2) Carefully determine the beginning and ending values of the axes. In this example, the units of measurement of the **x axis will be percents of birds from the original flock** beginning at 0 and ending at 100. The **y axis will represent the number of new birds** and begin with 0 and end with 25.

Step 3) Turn the table values into coordinates: (74, 5), (66, 6), (81, 8), (52, 11)… (38, 20).

Step 4) Graph the points from step 3.

Step 4) Done. Stand back and admire your work.

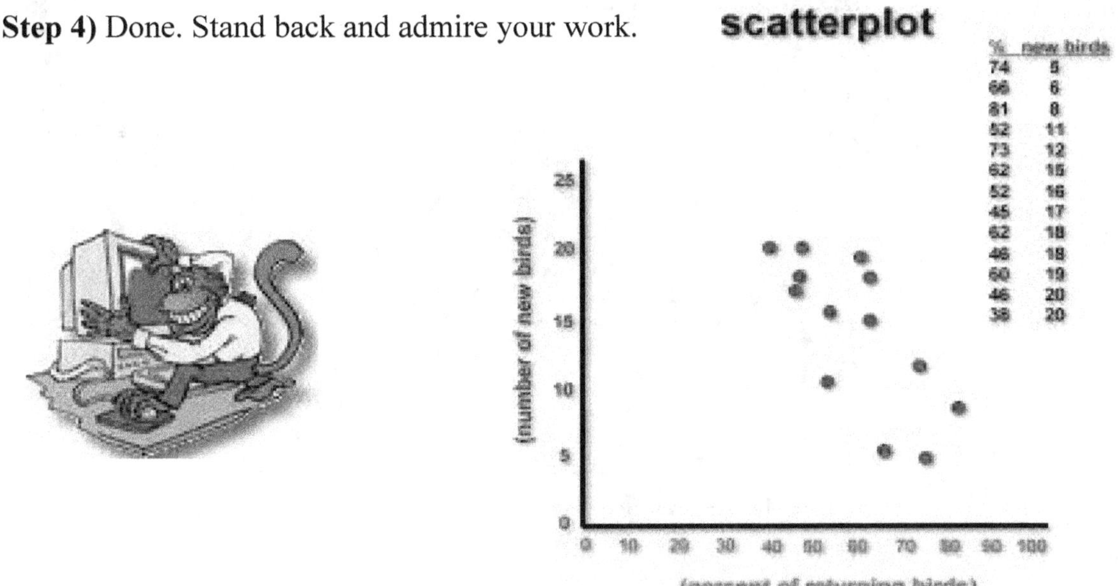

Correlation

The ***correlation*** measures the ***strength*** and ***direction*** of the linear relationship between two quantitative variables. **All correlations must be between -1 and 1, inclusive**.

correlation

The value of a correlation has to be between -1 and 1, inclusive.

If you have a value greater than 1, or less than -1, you have made an error somewhere.

Any correlation relatively close to 1 or -1 is considered a ***strong*** correlation. Any correlation that is close to 0 it is called a ***weak*** correlation. If the correlation is a negative number then it is called a ***negative*** correlation, and if it is a positive number it is called a ***positive*** correlation.

A **negative correlation** shows that when one variable increases in value, the other variable decreases, or vice versa. Another name for a negative correlation is an *inverse* variation.

A **positive correlation** is when one variable increases in value, the other variable increases as well. Another name for it is a *direct* variation.

Here is the correlation formula:

$$r = \frac{1}{n-1}\Sigma\left(\frac{x_i - \bar{X}}{s_x}\right)\left(\frac{y_i - \bar{Y}}{s_y}\right)$$

Are you afraid of these scary math formulas?

And here is how to interpret this formula:

$$r = \frac{1}{n-1}\Sigma\left(\frac{x_i - \bar{X}}{s_x}\right)\left(\frac{y_i - \bar{Y}}{s_y}\right)$$

correlation (r)

This means add them all up (Σ)

each measurement from the x variable (x_i)

mean of the x variable ($\bar{X}$)

each measurement from the y variable (y_i)

mean of the y variable ($\bar{Y}$)

sample size (n)

standard deviation of the first variable (s_x)

standard deviation of the second variable (s_y)

Example 4.3: Migratory sparrows were counted for thirteen years. The percent of birds from the original flock was recorded along with the number of new sparrows that had recently joined the flock. Find the ***correlation*** between the percent of birds from the original flock and the number of new birds.

percent	new birds
74	5
66	6
81	8
52	11
73	12
62	15
52	16
45	17
62	18
46	18
60	19
46	20
38	20

Step 1) Get your variables straight! In this problem we will call the percent of the original flock as the variable x, and the new birds as the variable y.

Step 2) Find the mean of the x variable and the y variable:
$\overline{X}$ = **58.23** and $\overline{Y}$ = **14.23**

Step 3) Find the standard deviation of the x variable and the y variable: s_x **= 58.23 and** s_y = **14.23**

Step 4) Time to use the correlation formula with lots and lots of calculating. Let's start with the x variable first.

x_i		$\bar{x}$				S_x		
74	-	58.23	≈	15.77	÷	13.03	≈	1.21
66	-	58.23	≈	7.77	÷	13.03	≈	0.6
81	-	58.23	≈	22.77	÷	13.03	≈	1.75
52	-	58.23	≈	-6.23	÷	13.03	≈	-0.48
73	-	58.23	≈	14.77	÷	13.03	≈	1.13
62	-	58.23	≈	3.77	÷	13.03	≈	0.29
52	-	58.23	≈	-6.23	÷	13.03	≈	-0.48
45	-	58.23	≈	-13.23	÷	13.03	≈	-1.02
62	-	58.23	≈	3.77	÷	13.03	≈	0.29
46	-	58.23	≈	-12.23	÷	13.03	≈	-0.94
60	-	58.23	≈	1.77	÷	13.03	≈	0.14
46	-	58.23	≈	-12.23	÷	13.03	≈	-0.94
38	-	58.23	≈	-20.23	÷	13.03	≈	-1.55

Now the variable y.

y_i		$\bar{y}$				S_y		
5	-	14.23	≈	-9.23	÷	5.29	≈	-1.74
6	-	14.23	≈	-8.23	÷	5.29	≈	-1.56
8	-	14.23	≈	-6.23	÷	5.29	≈	-1.18
11	-	14.23	≈	-3.23	÷	5.29	≈	-0.61
12	-	14.23	≈	-2.23	÷	5.29	≈	-0.42
15	-	14.23	≈	0.77	÷	5.29	≈	0.15
16	-	14.23	≈	1.77	÷	5.29	≈	0.33
17	-	14.23	≈	2.77	÷	5.29	≈	0.52
18	-	14.23	≈	3.77	÷	5.29	≈	0.71
18	-	14.23	≈	3.77	÷	5.29	≈	0.71
19	-	14.23	≈	4.77	÷	5.29	≈	0.9
20	-	14.23	≈	5.77	÷	5.29	≈	1.09
20	-	14.23	≈	5.77	÷	5.29	≈	1.09

Step 5) Multiply the last column of the x data with the last column of the y data:

$\frac{x_i - \bar{x}}{S_x}$		$\frac{y_i - \bar{y}}{S_y}$		
1.21	x	-1.74	≈	-2.11
0.6	x	-1.56	≈	-0.93
1.75	x	-1.18	≈	-2.06
-0.48	x	-0.61	≈	0.29
1.13	x	-0.42	≈	-0.48
0.29	x	0.15	≈	0.04
-0.48	x	0.33	≈	-0.16
-1.02	x	0.52	≈	-0.53
0.29	x	0.71	≈	0.21
-0.94	x	0.71	≈	-0.67
0.14	x	0.9	≈	0.12
-0.94	x	1.09	≈	-1.02
-1.55	x	1.09	≈	-1.69

Step 6) Add up the column of products between the x data and the y data: **-8.99**

Step 7) Divide the sum from step 6 by one less than the sample size (***n-1***): -8.99/(13 - 1) = **-0.749.** This is the **correlation!**

Step 8) Did mgz lie about all of the calculations?

So, the correlation between the percent of birds from the original flock and the number of new birds is -0.749. This correlation is negative and relatively strong (it is sort of close to -1). This correlation states that the higher the percentage of birds from the original flock then the less number of new birds joining the flock. Does it make sense? Yes it does.

Lines of regression

After you have collected two sets of measurements, or observations, on two different variables for each individual in your study, and after you have created the scatterplot that illustrates these data points, it is time to find the best straight line that best "***fits***" all of the points on the scatterplot.

These best fit lines are called ***lines of regression***.

A regression line can be linear (a straight line) and that fact allows us to use the **y-intercept/slope equation**, commonly called **y = mx + b**, from algebra. The regression line describes how the response variable (the y from the ***y*** = ***mx*** + ***b*** equation) changes as the explanatory variable (the x from the linear equation) changes.

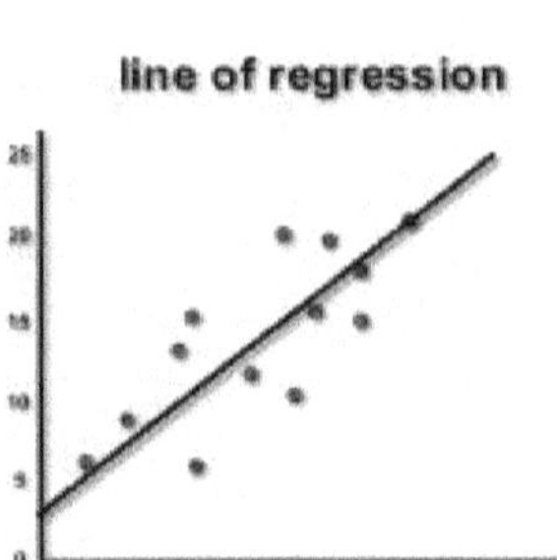

The proper method for writing a regression line is as follows: ***y-hat*** = ***a*** + ***bx***. The y-intercept is ***a*** and the slope is ***b***.

If the slope of the line is a positive number, then the relationship between the variables is called positive, and if the slope is a negative number then the relationship is called negative.

The principal use of a regression line is that it allows us to predict the value of the response variable for any given value of the explanatory variable.

Least-squares regression line

The king of regression lines is the ***least-squares regression line***. Using the following formula, this line minimizes the sum of the squared vertical distances between the observed points (the "real" points on the scatterplot) and the points on the line. Remember that this is a "best fit" line and will not exactly predict values. The formula is kind of long, but here it is:

$$\hat{y} = a + bx$$

(y intercept) $a = \bar{y} - b\bar{x}$

(slope) $b = r(sy/sx)$

($\hat{y}$ is the predicted value of the y variable)

Vocabulary

Least-squares regression line is a straight line that cuts through a set of points on a scatterplot that best fits all of the data points.

The **y-hat** is the ***predicted*** y value (predicted from the formula and therefore on the best-fit line), **r** is the correlation (yikes!), s_y is the standard deviation for the y variable, s_x is the standard deviation of the x variable, **y-bar** is the mean of the y values and **x-bar** is the mean of the x values.

The ***actual vertical distance between*** (on the y axis only) an observed point (one of the "real" scatterplot points) and its corresponding predicted point on the line of regression is called the ***residual***.

The residual = observed (real) y - predicted y .... Residual = y – y-hat

Example 4.4: Migratory sparrows were counted for thirteen years. The percent of birds from the original flock was recorded along with the number of new sparrows that had recently joined the flock. Find the equation for the ***least-squares regression line*** between the percent of birds from the original flock and the number of new birds.

percent	new birds
74	5
66	6
81	8
52	11
73	12
62	15
52	16
45	17
62	18
46	18
60	19
46	20
38	20

Step 1) To find the equation of the least-squares line of regression, we need to find the means and standard deviations from both the x and y variables, and the correlation between the two variables as well.

$$s_x = 13.03$$
$$s_y = 5.29$$
$$r = -.75$$
$$\bar{x} = 58.23$$
$$\bar{y} = 14.23$$

Step 2) Find the ***b*** (slope) first because you need to know that value in order to solve for the ***a*** (y intercept).

$b = r(s_y / s_x)$ ... $b = -.75(5.29/13.03)$... $b = -.304$

Step 3) Now find ***a*** (y intercept).

$a = \text{y-bar} - b(\text{x-bar})$ ... $a = 14.23 - (-.304)(58.23)$ ...
$a = 31.93$

Step 4) Plug in the ***a*** and ***b*** values into the least-squares line of regression formula.

y-hat* = *a* + *bx ... ***y-hat* = 31.93 + (-.304)*x***

Step 5) So the least-squares line of regression for this problem is ***y-hat* = 31.93 - .304*x***. This is stating that every time the ***x*** variable changes by a positive single unit, then the value of the y variable will decrease by 0.304. When the percentage of birds from the original flock goes up by 1%, then the number of new birds joining the flock goes down by 0.304. In other words, when the percentage of birds from the original flock goes up by 3%, then the number of new birds joining the flock goes down by 1 bird.

Chapter 4 Review

A ***correlation*** is a statistical tool that measures the relationship between two quantitative variables. In statistics we use the correlation to investigate if one variable is influencing, or *controlling*, the other variable. Remember that *correlation does not prove causality*!

Chapter Review

A variable that *controls* the value of another variable is called the ***explanatory*** variable because it ***explains*** what is happening to the other variable. The variable that is being controlled is called the ***response*** variable because it is simply ***responding*** to the explanatory variable.

A ***scatterplot*** is a graph of recorded data on a Cartesian plane of the relationship between two variables. Usually the explanatory variable is plotted on the x axis and the response variable is plotted on the y axis.

The correlation measures the ***strength*** and ***direction*** of the linear relationship between two quantitative variables. All correlations must be between -1 and 1, inclusive.

A strong correlation is relatively close to 1 or -1. A ***weak*** correlation is close to 0. If the correlation is a negative number then it is called a ***negative*** correlation, and if it is a positive number it is called a ***positive*** correlation.

A ***linear regression line*** is a straight line that passes though the points of a scatterplot that best fits the overall data. The regression line describes how the response variable (y) changes as the explanatory variable (x) changes.

The ***least-squares regression line*** is a line that minimizes the sum of the squared vertical distances between the observed point (the "real" point on the scatterplot) and the point on the line (residuals).

Guided homework Chapter 4: Correlations

Farmer Joe wanted to figure out how the height of his corn was related to how much water it received on a daily basis. He created six different plots (of the same size) and planted his corn. He watered them according to the data in the chart below, and recorded their heights after 2 months.

mathguyzero

watch the online video explanations

Amount of water (liters)	2.2	2.8	3.4	4	4.6	5.2
Height of plant (cms)	212	214	217	219	220	222

Use the data from the table above to answer the following questions.

1. Which is the response variable and which is the explanatory variable?

2. What is the correlation?

3. Is there a strong or weak relation between the variables?

4. A positive or negative correlation?

5. What is the equation for the least-squares line of regression?

6. Use the regression line equation to predict how high the corn should be if it received 8 liters of water daily.

Chapter 4 Quiz

1) A pharmacologist invents a new "algebra pill" that she states helps students learn algebra. In her experiment, she gave an algebra pre-test to 6 students, and then she administered different doses (in mgs) of the new learning drug to each of the subjects. Then she gave a similar post test. The following table lists the results of the dosages and the difference in their corresponding test scores.

Dosage	Score
0.15	6
0.2	4
0.25	8
0.3	7
0.35	6
0.4	9

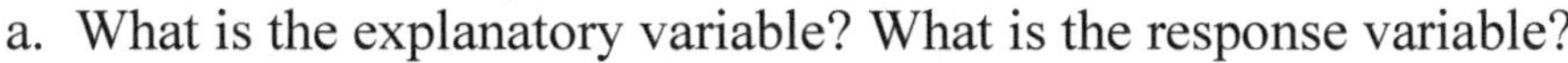
a. What is the explanatory variable? What is the response variable?

b. What is the correlation?

c. Is the correlation weak or strong? Positive or negative?

d. Make a scatterplot of the data.

e. What is the least-squares regression line?

f. According to the least-squares regression line, what is the predicted difference in pre and post test scores if the subject was given an algebra pill dosage of .85 mgs?

2) According to the scatterplot below, what is the residual at the data point of (0.2, 2)?

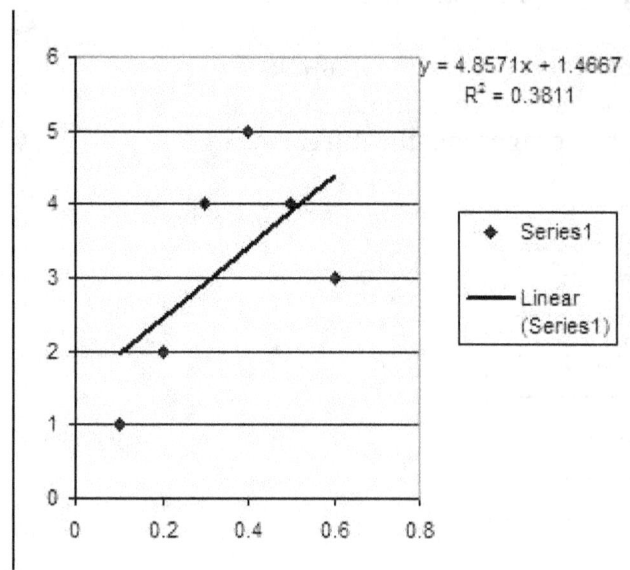

3) Which of the following could never be a correlation?

a. -1.67

b. -.09

c. .74

d. .65

e. -.87

4) True or false?

A strong correlation proves beyond a shadow of a doubt that one variable is "causing" the other variable.

5) What is the term that describes an unseen, untested, variable that may be affecting both variables in a correlation?

a. a response variable

b. a short variable

c. a blind variable

d. a lurking variable

e. a wide variable

6) True or false?

The correlation is the slope of the least-squares line of regression.

Ch 5: General Probabilities

Chapter
5

General Probabilities

If the amount of luck a person has could be quantified (somehow – what would you consider good luck?), then the luck *measurements* from a group of random people should fit a normal distribution. That means that half of the people would have less than the average amount of luck (unlucky) while the other half would have more than the average amount. Lucky people! Do you feel lucky…punk? (- old movie reference)

Luck is simply an event with a positive outcome for whoever (or whatever) is experiencing the event. Another word for this event-and-outcome experience is called ***chance***.

Statisticians call chance ***probability***. ***Probability does not predict individual outcomes,*** but rather it illustrates long-term trends and predicts the overall pattern of the outcomes if the event were to be repeated under the exact same conditions many times.

What's in this chapter

- Events and outcomes
- Probability models
- Probability rules
- Conditional probability
- Discrete and continuous probabilities

In this chapter, we will investigate general probabilities:

- Events and outcomes

- Probability models

- Probability rules

- Conditional probability

- Discrete and continuous probabilities

Events and outcomes

In probability, an ***event*** is a trial that has more than one possible outcome. Events can be the flipping of a coin, being dealt some playing cards, picking a green M&M out of a bag of M&Ms, etc.

An ***outcome*** is the result of each event or trial. A coin coming up tails, being dealt three kings, or picking a green M&M are all outcomes. Outcomes can also be ***not*** being dealt three kings or ***not*** picking the green M&M.

In order to use the probability models covered in this chapter, each outcome must be a random outcome and have the exact same chances of occurring as any other outcome. Rolling a single six-sided die is a good example of all outcomes having the same chance of occurring because each of the six numbers has the same chance of showing, unless the die is weighted on one side to make it show a specific number more frequently. That is called cheating. For shame!

Random outcomes follow long-term patterns. Take the six-sided die for example: if you rolled it a million times you would expect to get the number 4 approximately 1/6th of the time. In fact, you would expect that each of the six numbers to show up 1/6th of the time, but never exactly 1/6th of the time. This expectation of results from many repeated events is called the ***probability of the outcome occurring***.

*Note: Remember that probability can not predict individual outcomes, only long-term trends!

Probability Models

Probability models ***will not predict the outcome from a single event***. They show the overall pattern, or trend, of the outcomes if the event is repeated many many times.

Probability models

A ***probability model*** describes how the outcomes from events will behave in the long run. Probability models all work the same: they show the ratio of a single outcome to the total number of all possible outcomes from the event. The set of all possible outcomes from an event is called the ***sample space S***.

Example 5.1: What is the probability of picking the Jack of Hearts from a 52 playing card deck on one draw?

The desired outcome is picking the Jack of Hearts, of which there is only 1. The sample space S is all possible outcomes:

every card in the deck is a possible outcome (52 different possible outcomes). So the probability of picking the Jack of Hearts on a single draw is 1 out of 52, or 1/52, or as a decimal, close to 0.019.

Example 5. 2: What is the probability of picking any queen on a single draw from a 52 playing card deck?

The desired outcome is any queen, of which there are 4 in the deck. The sample space S is 52 (all of the cards). So the probability of picking any queen on a single draw is 4 out of 52, or 4/52, or 1/13, or as a decimal, close to 0.077.

Sample space S can be simple or complex. The sample space S of a single coin toss has only two different outcomes: heads or tails, H or T. But the sample space S of three coin tosses has 8 possible outcomes: HHH, HHT, HTH, HTT, THH, THT, TTH and TTT.

*Note: The sum of all probabilities of all possible outcomes for any single event is equal to 1.

Vocabulary

Sample space S
The sample space S is a list of all possible outcomes of a single event.

Example
The sample space S of a single coin toss: H and T (either heads or tails)

Example
The sample space S of a single six-sided die: 1, 2, 3, 4, 5, 6.

The formula to find the number of all possible outcomes in a sample space S is:

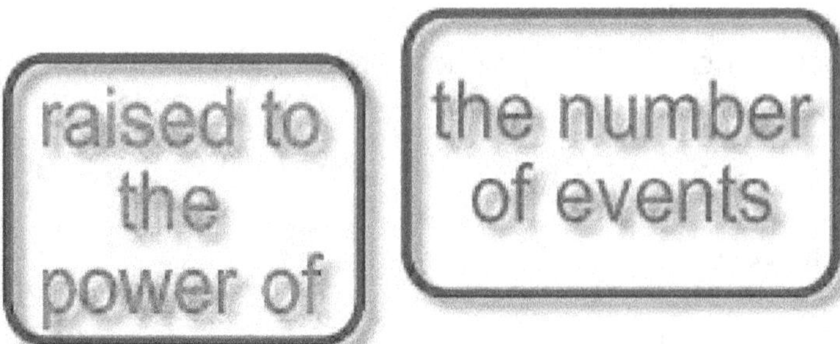

the number of different outcomes per event

Example 5.3: The number of possible outcomes from a sample space S of a single roll with two six-sided dice is 36.

Sample Space (S) of two six-sided dice

first die	second die	first die	second die	first die	second die
1	1	2	1	3	1
1	2	2	2	3	2
1	3	2	3	3	3
1	4	2	4	3	4
1	5	2	5	3	5
1	6	2	6	3	6

first die	second die	first die	second die	first die	second die
4	1	5	1	6	1
4	2	5	2	6	2
4	3	5	3	6	3
4	4	5	4	6	4
4	5	5	5	6	5
4	6	5	6	6	6

Example 5.4: The total number of possible outcomes in a sample space for the wonderful game of Yahtzee where five six-sided dice are tossed at once is 6 (the number of different outcomes per event) **raised** to the 5th power (because there are 5 dice being rolled): 6^5 = **7776**. (7776 different possible outcomes!)

Example 5.5: The sample space for 10 coin tosses is 2 (the number of different outcomes per event: H or T) **raised** to the 10th power (10 coin tosses): 2^{10} = **1024**.

Probability rules

Some probability models are extremely complex, especially if one outcome influences other outcomes. But there are ***five basic rules*** that apply to all probability models.

*Note: ***P(A)*** is the way that statisticians write ***the probability of event A occurring***.

<u>Rule 1:</u> All probability has to be a number between, or equal to, ***0 and 1: $0 \leq P(A) \leq 1$***. Probabilities can be written as fractions (ratios) but are normally written as percents or decimals. Remember that there is no such animal as a ***negative probability***!

P(A)

P(A) is a shortcut to writing "the probability of event A occurring".

<u>Rule 2:</u> Since all probabilities have to be between 0 and 1 inclusive, the sample space S is always the denominator of the fraction, or ratio, that defines the probability, and then ***the fraction is usually converted to an equivalent percent or decimal with four significant place values***.

<u>Rule 3:</u> If two separate events, A and B, can never occur at the same time, they are called ***disjoint events***. Getting both a head and a tail on a single toss would never happen in this universe and is a good example of disjoint events. ***If events A and B are disjoint***, then the probability of ***P(A or B) = P(A) + P(B)***. **<u>(In probability, "or" means to add.)</u>**

<u>Rule 4:</u> If A and B are disjoint events, then the probability of ***P(A + B) = P(A)* x *P(B)*** . **<u>(In probability, "and" means to multiply.)</u>**

<u>Rule 5:</u> For any event A with a probability of occurring of P(A), then the probability of event A ***not*** occurring, ***P(A does not occur) = 1 - P(A).***

Example 5.6: What are the chances of getting three tails from three coin tosses?

The sample space S of three coin tosses is $2^3 = 8$, and there is only 1 way (outcome) that will result in three tails (TTT). So the probability of getting three tails out of three coin tosses is 1/8. Now change the 1/8 into a decimal or a percent. 1/8 = .1250 = **12.50%**

Example 5.7: What are the chances of ***not*** getting three tails from three coin tosses?

The sample space S of three coin tosses is $2^3 = 8$, and there is only 1 way (outcome) that will result in three tails (TTT). So the probability of ***not*** getting three tails out of three coin tosses is 1 - 1/8. Now change the 7/8 into a decimal or a percent. 7/8 = .8750 = **87.50%**

Example 5.8: What is the probability of picking the Jack of Hearts ***or*** any queen on a single draw?

The chance of picking the Jack of Hearts is 1/52 and the chance of getting any queen is 4/52. So the probability of picking the Jack of Hearts ***or*** any queen on a single draw:

1/52 + 4/52 = 5/52 = .0962 = 9.62%

Example 5.9: What is the probability of getting a 4 on one die ***and*** a 2 on another die in a single roll of two six-sided dice?

The probability of getting a 4 or a 2 on the first dice is 1/6 + 1/6 = 1/3, and the probability of getting a 2 or a 4 on the second dice is also 1/6 + 1/6 = 1/3. So the probability of getting a 4 **and** a 2 on a single roll of two six-sided dice is:

(1/3) x (1/3) = 1/9 = .1111 = 11.11%

Note: In logic and probabilities, ***or* is the same as ***union*** while ***and*** is the same as ***intersect***.

Conditional probability – General Addition Rule

Remember rule 3 from the general probability rules? Here it is again: ***If two separate events, A and B, can never occur at the same time, they are called disjoint events, or independent events***. A good example of disjoint events would be to get both a head and a tail on a single coin toss. That would never happen in this universe.

If events A and B are disjoint, then the probability of P(A or B) = P(A) + P(B). (In probability "or" means to add.)

But sometimes two different events are ***not disjoint*** and they can both occur at the same time. Example: event A = having blue eyes and event B = being a psychology major. It is indeed possible to find many blue-eyed psych majors. When one outcome is necessary, or needed, before another event's outcome can be calculated, then these probabilities are called ***conditional probabilities***.

For these overlapping outcomes, or occurrences, we use the ***General Addition Rule*** for any two non-disjoint events. Notice the graphic on the next page that events A and B overlap in the intersection of the two sets.

If we were counting the outcomes of event A and event B separately, we would count the outcomes in the intersection of A and B twice. Wrong!

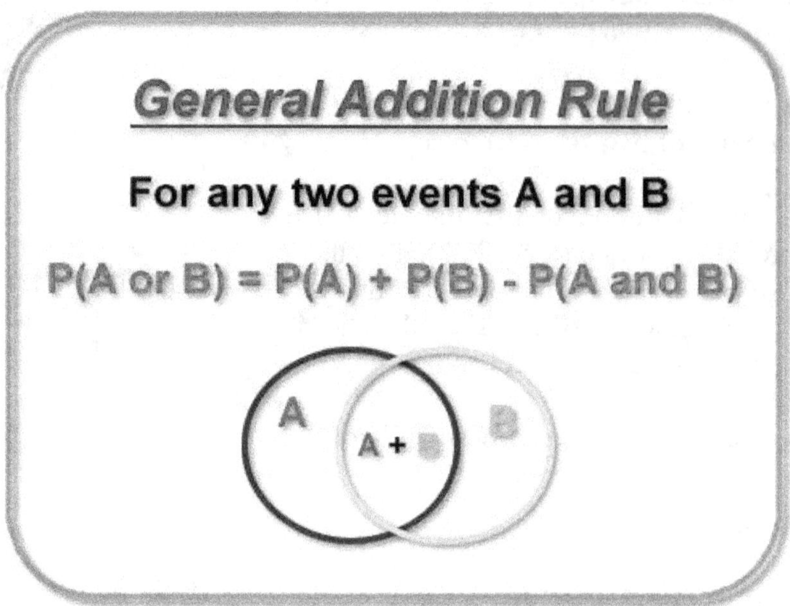

Example 5.10: Desserts at local elementary schools were classified and counted. The two most popular desserts were ice cream and pudding and the two favorite flavors were chocolate and vanilla, naturally. The distribution is listed in the table below. A total of 32% of the favorite flavors were vanilla and 53% were chocolate, and 60% of the most popular desserts were ice cream and 25% were pudding.

Desserts	**Vanilla**	**Chocolate**	Total
Ice cream	20%	40%	**60%**
Pudding	12%	13%	25%
Total	**32%**	**53%**	

Question 1) What is the probability that one dessert chosen at random would be vanilla flavored ***or*** ice cream?

Answer 1) P(vanilla ***or*** ice cream) = P(vanilla) + P(ice cream) - P(vanilla ice cream) = 32% + 60% - 20% = **72%**

Question 2) What is the probability that one dessert chosen at random would be chocolate flavored ***or*** pudding?

Answer 2) P(chocolate ***or*** pudding) = P(chocolate) + P(pudding) - P(chocolate pudding) = 53% + 25% - 13% = **65%**

Conditional probability - B given A

What if the probability of event B depended on event A happening first? These intertwined probabilities are also called ***conditional probabilities***.

For these events "**A before event B**" probabilities we use the ***Conditional Probability Rule*** for any two events. Please notice that P(A) has to be a non-zero probability. Do you know why?

Conditional Probability Rule

The probability of B occurring after A has already occurred first (B given A) is

$$P(B \mid A) = \frac{P(A \text{ and } B)}{P(A)}$$

Example 5.11: Desserts at local elementary schools were classified and counted. The two most popular desserts were ice cream and pudding and the two favorite flavors were chocolate and vanilla. The distribution is listed in the table. 32% of the

favorite desserts were vanilla flavored and 53% were chocolate, and 60% were ice cream and 25% were pudding.

Desserts	Vanilla	Chocolate	Total
Ice cream	20%	40%	**60%**
Pudding	12%	13%	25%
Total	32%	53%	

Question 1) What is the conditional probability that one dessert chosen at random will be chocolate given that it is ice cream?

Answer 1) P(chocolate | ice cream) = P(chocolate and ice cream) divided by P(ice cream) = 40% / 60% = **66.7%**

Question 2) What is the conditional probability that one dessert chosen at random will be pudding given that it is vanilla?

Answer 2) P(pudding | vanilla) = P(pudding + vanilla) divided by P(vanilla) = 20% / 32% = **62.5%**

Conditional probability - General Multiplication Rule

All ***conditional probabilities*** are normally stated in the original data and with this given information we can use the ***General Multiplication Rule*** for finding the probability that both events A and B occurring at the same time.

General Multiplication Rule

The probability that both events A and B will occur together is

P(A and B) = P(A) x P(B | A)

Example 5.12: The National Internet Research Institute states that 68% of American males between the ages of 20 - 28 have taken an online course and that 23% of them have watched online educational videos on YouTube. We will let event A = taken an online course, and event B = watched online educational videos.

Question 1) What percent of American males between the ages of 20 - 28 have taken a online course ***and*** watched educational videos on YouTube?

Answer 1) P(online course and videos) = P(online course) x P(videos | online course)

P(A) = 68% and P(B | A) = 23%, therefore the probability that any American male between the ages of 20 - 28 who have taken an online class ***and*** have watched educational videos on YouTube is 68% x 23% = **15.64%.**

Discrete and continuous probability models

All probability models with a ***finite*** sample space S are called ***discrete probability models.*** A total list of possible outcomes is doable with discrete models, even if it is a long list. You could write out all of the possible outcomes if you flipped a coin 100 times if you wanted to (that would be 1,267,650,600,228,229,401,496,703,205,376 or roughly 1.3 nonillions different combinations of heads and/or tails), but don't do that! Use the math instead. That's why it was *discovered* in the first place!

Discrete probabilities can be written in table form, which we can use to answer questions about the probabilities of each of the outcomes.

Example 5.13: Here is an example of a frequency distribution of the colors of automobiles:

Discrete probability table

Car color	Silver	White	Red	Black	Gray	Brown	Blue	Green
Probability	20%	18%	13%	12%	12%	9%	9%	6%

Question 1) What is the probability that a randomly chosen car will be red?

Answer 1) **13%**. Easy, cheesy, bright and breezy.

Question 2) What is the probability that a randomly chosen car will be black ***<u>or</u>*** white?

Answer 2) P(A or B) = P(A) + P(B) = 12% + 18% = **30%**

Question 3) What is the probability that two randomly chosen cars will be green ***<u>and</u>*** blue?

Answer 3) P(A and B) = P(A) x P(B) = 6% x 9% = **0.54%**

Question 4) What is the probability of a randomly chosen car will be purple?

Answer 4) Notice that purple is not on the table. So, sum up the probabilities listed and subtract that sum from 1 or 100%: 100% - (20% + 18% + 13% + 12% + 12% + 9% + 9% + 6%) = 1%. Therefore, the probability of a randomly chosen car will be purple are less than, or equal to, 1%... **P(A) ≤ 1%**

But not all sample spaces are finite. A good example of an infinite sample space would be if you were to randomly choose any decimal between 0 and 1. The sample space S would be all of the decimals between 0 and 1. How many different decimals are there between 0 and 1? A bunch! A great big bunch! An infinite number of infinities! All creatures that have ever lived in the entire universe writing decimals at the speed of light would never be able to write out all of the decimals that exist between 0 and 1. Cool stuff those infinities, eh?

All probability models with an infinite sample space are called ***continuous probability models***. In other words, continuous probability models are those with measurements that can have a partial unit, such as a tenth of a second, or a fifth of a degree. Since we can't use infinity as a denominator of a fraction, we assign these probabilities with a density curve, usually with a normal distribution curve or a software program.

Chapter 5 Review

Chance, or ***probability***, predicts the ***overall pattern*** of the outcomes of an event, or trail, ***if the event were to be repeated under the exact same conditions many times.*** Probability does not predict individual outcomes, but rather it illustrates long-term trends.

A ***probability model*** describes how the outcomes from events will behave in the long run. Probability models compare the ratio of a single outcome occurring to the total number of all possible outcomes from an event. The set of all possible outcomes from an event is called the ***sample space S***.

General probability rules:
Rule 1: ***$0 \leq P(A) \leq 1$***. There is no such thing as a negative probability!
Rule 2: The sample space S is always the denominator of the fraction, or ratio, that defines the probability, and the chance of the specific outcome is the numerator.
Rule 3: If two ***disjoint*** events, A and B, then ***P(A or B) = P(A) + P(B)***. (In probability, "***or***" means to add.)
Rule 4: If two ***disjoint*** events, A and B, then the probability of ***P(A + B) = P(A) x P(B)***. (In probability, "***and***" means to multiply.)
Rule 5: For any event A with a probability of occurring of P(A), then the probability of event A ***not*** occurring, ***P(A does not occur) = 1 - P(A).***

Sample Random Space (SRS)

An SRS is a randomly chosen sample from a population.

An SRS can also be called a "simple" random space. The two definitions mean the same thing: no bias!

Finding the chances of an outcome from an event on the condition that another outcome has occurred first is called a ***conditional probability***. For these overlapping outcomes, or occurrences, we use the ***General Addition Rule*** for any two non-disjoint events:
P(A or B) = P(A) + P(B) – P(A and B).

For "**A before event B**" probabilities we use the ***Conditional Probability Rule*** for any two events: ***P(B | A) = (P(A and B))/P(A)***

The ***General Multiplication Rule*** for finding the probability that both events A and B occurring at the same time: ***P(A and B) = P(A) x P(B | A)***

All probability models with a ***finite*** sample space S are called ***discrete probability models.*** All probability models with an ***infinite*** sample space are called ***continuous probability models***.

Guided homework Chapter 5: General Probabilities

Cars were randomly selected and the following probabilities reflect their different colors.

Color	Black	White	Red	Blue	Grey/Silver	Green
Probability	0.08	0.22	0.18	0.16	0.21	0.13

1. You are sitting on a lawn chair in your front yard. What is the probability that the next car that drives by:

- Will be red?
- Will be grey or white?
- Will not be black or white?
- Will be purple?

2. Shoes were randomly selected and the following probabilities reflect their different colors.

Shoe Color	Black	White	Red	Blue	Grey/Silver	Brown
Probability	0.12	0.32	0.18	0.22	0.12	0.13

What is wrong about this probability table?

3. What are the chances of rolling an 18 with three dice in a single roll?

4. What are your chances of rolling a 2, 3 or 12 with two dice in a single roll?

5. What are the chances of rolling a 7 or an 11 with two dice in a single roll?

6. What would the SRS be of you rolled five dice at once, like in the wonderful game of YAHTZEE?

Guided homework Chapter 5: General Probabilities

7. Karnak is playing five card stud poker. In stud poker, each player is dealt five cards face up. There are 52 cards in a standard deck.

Event A = Dealt the Jack of Hearts
Event B = Dealt the Jack of Spades
Event C = Dealt any King

a) What is the P(A) happening for Karnak in a single hand?

b) What is the P(A or B) happening for Karnak in a single hand?

c) What is the probability of Event C happening for Karnak in a single hand?

d) What is the P(A), P(B), and P(C) happening for Karnak in a single hand?

e) What is the probability that Karnak will not be dealt any king, the Jack of Spades or the Jack of Hearts in single hand?

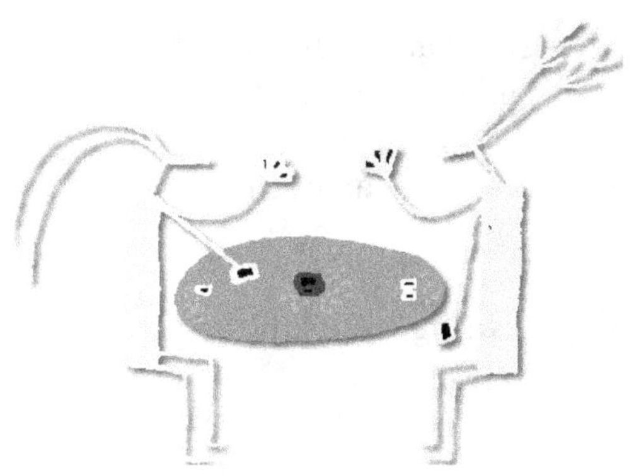

Chapter 5 Quiz

1) How many outcomes of a single roll of 5 six-sided dice are there?

2) True or false? - Statistics can be used to predict a single outcome.

3) The following probability table represents the favorite baseball teams of 1,000 randomly chosen Californians.

Dodgers	Padres	A's	Giants	Angels
0.23	**0.12**	**0.15**	**0.21**	**0.16**

a) What is the probability that one of the randomly chosen Californian's favorite team will be the Oakland A's?

b) What is the probability that one of the randomly chosen Californian's favorite team will be the Padres and another randomly chosen person is a Giant fan?

c) What is the probability that one of the randomly chosen Californian's favorite team will be the New York Yankees?

d) What is the probability that one of the randomly chosen Californian's favorite team will ***not*** be the Giants or the Padres?

e) What is the probability that one of the randomly chosen Californian's favorite team will be the Dodgers or the Angels?

4) True or false? - The sum of all probabilities of all outcomes from a single event is equal to 1.

5) A tetrahedron is a four-sided object. Let's pretend that you have 4 of these four-sided dice with sides numbering 1, 2, 3 and 4.

a) What is the probability that on a single roll you will get an even number?

b) What is the probability that on a single roll you will get a total of 5 or less?

6) What is the probability of picking the Queen of Spades or the Queen of Diamonds from a 52 card playing deck on a single pick?

7) If you rolled a six-sided die four times in a row and got a 5 each time, what is the probability that a 5 will come up on the next roll?

8) What is the probability of flipping a penny 8 times and getting 8 tails?

9) What is the probability of being dealt a royal flush (A, K, Q, J and 10) in spades in a game of five-card stud poker from a 52 card playing deck? (Players get 5 cards only, because they are muy macho!)

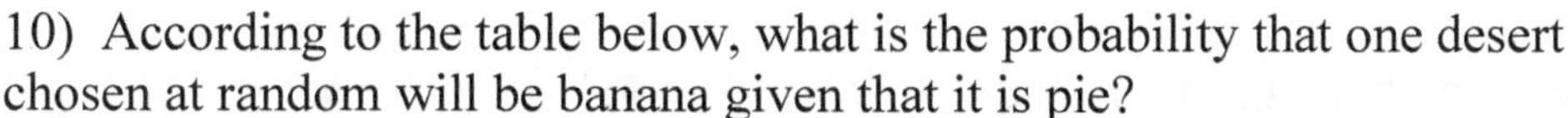
10) According to the table below, what is the probability that one desert chosen at random will be banana given that it is pie?

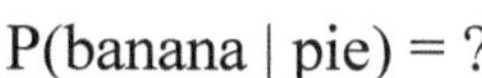
P(banana | pie) = ?

	Banana	**Peach**	**Total**
Pie	16%	12%	28%
Ice cream	14%	10%	24%
Total	30%	22%	

Midterm exam: Chapters 1 – 5

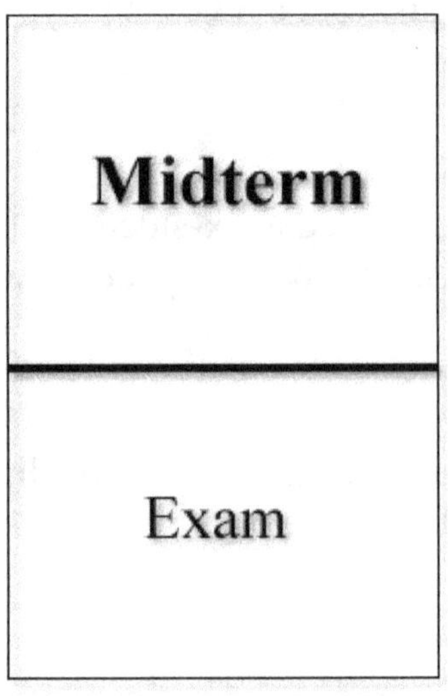

1. Fill in the blank with the best-fit choice below: Populations have parameters while samples have __________.

    a. measurements
    b. standard deviations
    c. variance
    d. statistics
    e. means

2. True or false? - If a person smokes or does not smoke is a quantitative variable.

3. A tetrahedron is a four-sided object. Let's pretend that you have 4 of these four-sided dice with sides numbering 1, 2, 3 and 4.

    a) What is the probability that on a single roll you will get a total of 15 or more?

    b) What is the probability that on a single roll you will get an odd number?

4. Which statement below is true about the boxplot?

    a. The median is around 50
    b. It is skewed to the right
    c. It is symmetrical
    d. It is skewed to the left
    e. Half of the data is between 50 and 100

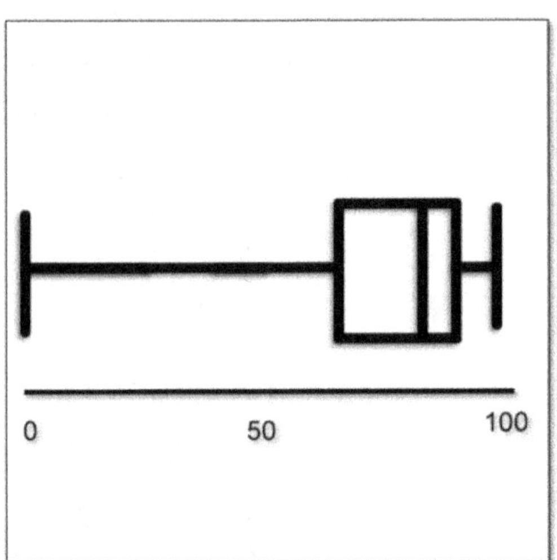

5. What is $P(z < -0.63)$?
    a) .5662
    b) -.5662
    c) .7457
    d) .9913
    e) .2643

6. True or false? – The number of miles that a person has to travel from home to school is a categorical variable.

Midterm

Exam

7. A pharmacologist invents a new "algebra pill" that she states helps students learn algebra (you wish!). In her experiment, she gave an algebra pre-test to 6 students, and then she administered different doses of the new learning drug to each of the subjects. Then she gave a similar post test. The following table lists the results of the dosages and the difference in their corresponding test scores.

Dosage	Score Difference
0.15	4
0.2	5
0.25	6
0.3	8
0.35	8
0.4	9

a) What is the correlation?

b) Which statement is true about the correlation?

a. strong and negative
b. parallel and weak
c. strong and positive
d. weak and positive
e. weak and negative

c) What is the equation of the least-squares line of regression?

d) Using the equation of the line of regression, what test score difference would be predicted with a dosage of 0.75?

8. From a normal distribution of test scores with a mean of 72 and a standard deviation of 12, N(72, 12), what would the standardized test score of 48 be?

Midterm

Exam

Exam

9. The following probability table represents the favorite baseball teams of 1,000 randomly chosen Californians.

Dodgers	Padres	A's	Giants	Angels
0.23	0.12	0.15	0.21	0.16

a) What is the probability that one of the randomly chosen Californian's favorite team will be the A's and another randomly chosen person is a Dodger fan?

b) What is the probability that one of the randomly chosen Californian's favorite team will be the Boston Red Sox?

c) What is the probability that one of the randomly chosen Californian's favorite team will be the Oakland A's or the Padres?

d) What is the probability that one of the randomly chosen Californian's favorite team will ***not*** be the Dodgers or the Angels?

10. True or false? - The sum of all probabilities of all outcomes from a single event **can be greater than 1**.

11. From a normal distribution of test scores with a mean of 72 and a standard deviation of 12, N(72, 12), what would the standardized test score of 97 be?

Midterm

Exam

12. The distribution of weights (in ounces) of "The Big Fudgy" ice cream bar is approximately normal with N(4.27, 0.33).

a) The smallest 0.15% of the ice cream bars will weigh under what weight?

b) Between what two weights do 68% of all the ice cream bars weigh?

c) The heaviest 2.5% of all of the ice cream bars will weigh over what weight?

13. The histogram below represents the grade point averages (GPA) of a high school's entire senior class.

a) According to the distribution table below, what percent of students have a GPA greater than 2.0?

b) According to the histogram below, how many students have a GPA between 2.0 and 4.0?

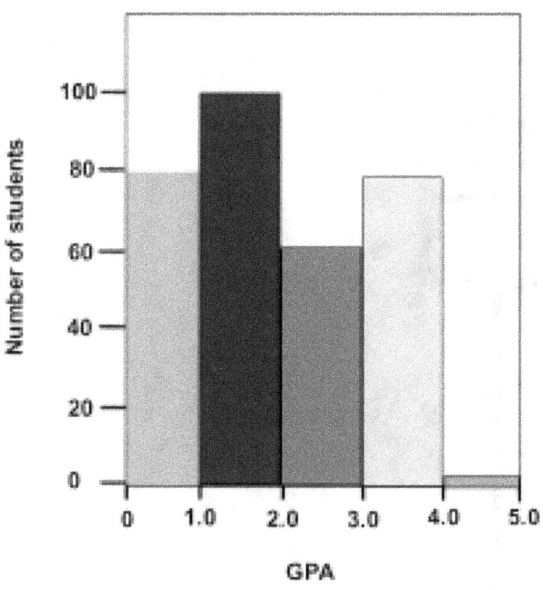

Midterm
Exam

14. The Calico Cat Company has a truckload of kittens with weights that are normally distributed. The mean weight of a kitten is 2.40 pounds with a standard deviation of 0.36 pounds. Little Virginia puts on a blindfold and reaches in and pulls out 1 cat at random. She takes the kitten home and names it "***Statistics Kitty***" and feeds it lots of tuna.

a) What is the probability that the kitten weighed less than 2 pounds when she first picked it from the truckload o' kittens?

b) The heaviest 5% of the kittens will weigh over what weight?

c) 50% of the kittens will weigh under what weight?

d) 95% of the kittens will weigh between what two weights?

e) The smallest 25% of the kittens will weigh under what weight?

f) What percent of the kittens will weigh between 2.5 and 2.9 lbs?

Midterm
Exam

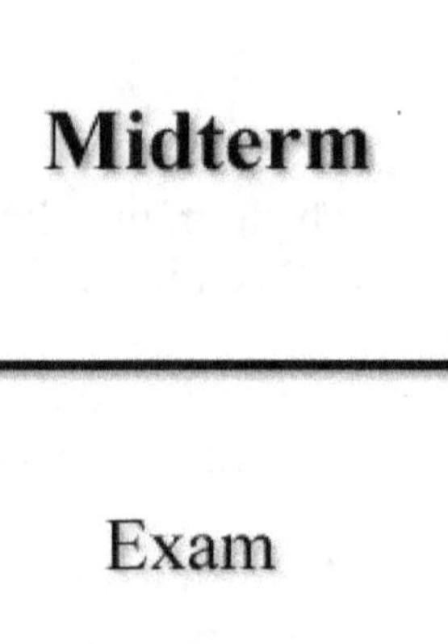

15. The boxplot below represents salaries of 400 randomly selected employed people.

a) According to the boxplot, the top 25% salaries are higher than what amount?

b) According to the boxplot, what percent of the people earn under $31,500?

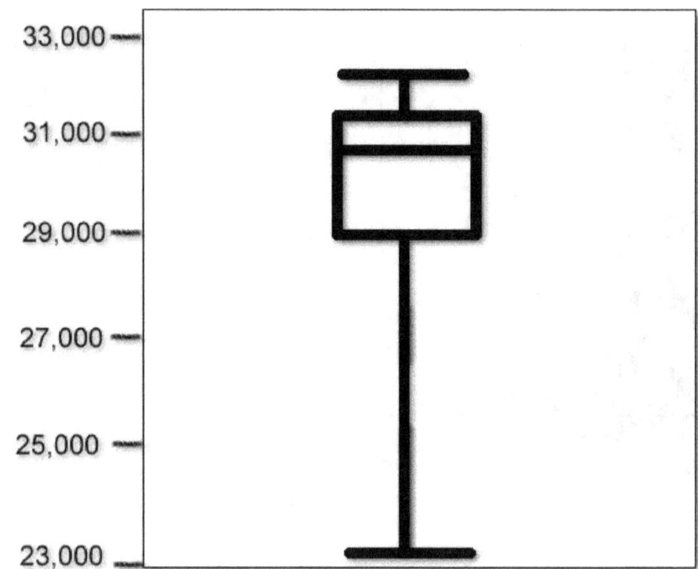

16. Which of the following numbers is not on the stemplot?

a) 87
b) 71
c) 69
d) 86
e) 901

stem	leaves
10	9
9	0, 1,
8	5, 6, 7
7	1, 7
6	4, 8 9

17. According to the table below, what is the probability that one desert chosen at random will be peach given that it is ice cream? P(peach | ice cream) = ?

	Banana	Peach	Total
Pie	16%	12%	28%
Ice Cream	14%	10%	24%
Total	30%	22%	

Midterm

Exam

18. There are _____ outcomes from a single roll of 4 six-sided dice.

19. Fill in the blank with the best-fit choice below: Parameters are to populations as _________ are to samples.

a. means
b. variance
c. statistics
d. measurements
e. standard deviations

20. What is the probability of flipping a quarter 6 times and getting 6 heads?

Midterm

Exam

21. If you rolled a single six-sided die four times in a row and got a 5 each time, what is the probability that a 5 will come up on the next roll?

22. True or false? - Statistics predicts long-term behavior but not a single outcome.

Midterm

Exam

23. What is the probability of picking the 2 of spades, or the 6 of diamonds or the Queen of Hearts from a 52 card playing deck on a single draw?

24. True or false? - A person's blood pressure is a quantitative variable.

25. From a normal distribution of test scores with a mean of 72 and a standard deviation of 12, N(72, 12), what would the standardized test score of 84 be?

26. If you standardized 600 test scores, what number would the sum of those z scores be close to?

a. 1
b. 100
c. 0
d. Not enough information to answer this question
e. 600

Midterm

Exam

27. What is $P(-2.03 < z < 1.79)$?

28. What is the probability of being dealt four 6s and the Jack of Hearts in a game of five-card stud poker from a 52 card playing deck? (Players get 5 cards only, all face up.)

29. According to the scatterplot below, what is the residual when x = 0.6?

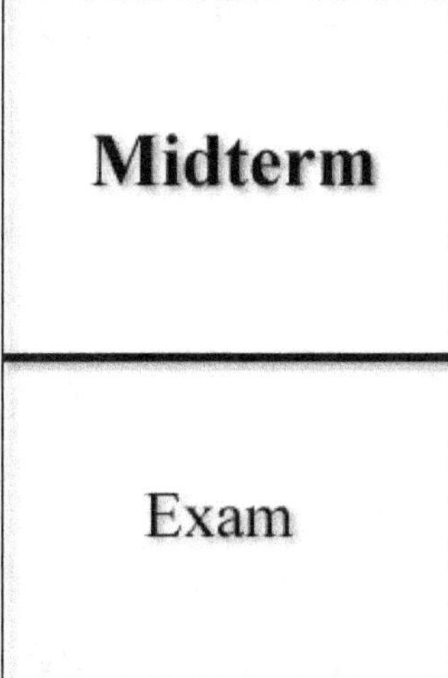

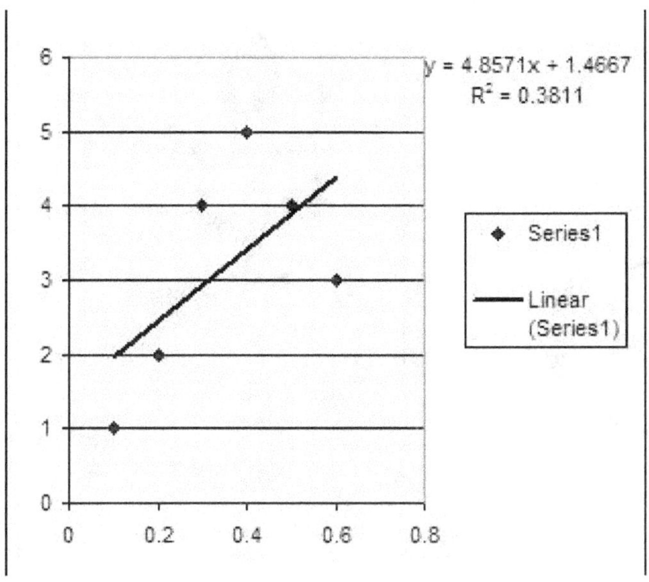

a. Close to 5.3
b. Close to -5.3
c. Close to 0.98
d. Close to -1.4
e. Close to 1.4

30. Tammy and Pammy are identical twin sisters with bright red hair and freckles... many many freckles.

Tammy scored 680 on the math section of the SAT test which has the distribution of N(560, 60). Pammy scored a 22 on the ACT test which has the following distribution (18, 2). Who had the higher standardized test score?

Ch 6: Binomial Probabilities

Chapter

6

Binomial Probabilities

Remember that an event has more than one possible outcome. In ***binomial probabilities***, the event must have only two possible outcomes, such as a head or tail from a coin toss, or if the ball goes through the hoop or not. These probability models are used by sports gamblers to calculate the betting odds.

We can use an algebraic approach to find the probabilities of any specific outcome that we choose. Remember that statistics shows a long-term pattern of behavior, not the outcome of any single event.

In this chapter, we will investigate ***binomial distributions*** and how we can predict long-term behavior:

- Binomial setting

- Binomial distributions

- Binomial probabilities

- Binomial coefficients

- Binomial means and standard deviations

- Normal approximations for binomials distributions

What's in this chapter

- Binomial setting
- Binomial distributions
- Binomial probabilities
- Binomial coefficients
- Binomial means and standard deviations
- Normal approximations for binomial distributions

Binomial settings

A new hair-growing formula that supposedly "can grow hair on a bowling ball" is tried on 500 big handsome bald math teachers with goatees. How many of them grew hair on their heads?

A botanist developed a new type of tomato plant that is supposed to grow a rainbow colored tomato. If 1,000 plants are grown, how many of them produced a rainbow tomato? (Color-liscious!)

The probability models for both of these situations are looking for the ***count*** of successful outcomes. The ***count*** is the ***number*** of successful outcomes. In our first example the ***count*** would be the number of guys that grew hair, and in the second the ***count*** would be the number of plants that produced a rainbow tomato.

In order to use the binomial probability formulas that calculate the count of successful outcomes, the data must fit the ***binomial setting***.

Here are the necessary conditions in order to use the mathematics behind finding binomial probabilities:

binomial setting

1) There is a large but finite sample size n

2) No outcome can influence any other outcome

3) Each outcome is either a "success" or a "failure"

4) The probability for success of any outcome is the same for every outcome

The coin toss is an excellent example of a binomial setting. A person can flip a coin as many times as he/she wants which satisfies rule 1; getting a head on the 23rd flip does not influence the outcome of the 24th flip,

Vocabulary

Trial
A trial examines the outcomes from many events.

Event
An *event* in the binomial setting has only two possible outcomes: "success" or "failure".

Count
The *count* is the number of successes in a single trial.
P(X = K)
(**X** means random variable and K is the specific number of successful outcomes.)

satisfying rule 2; a head can be a success or failure depending on what set of outcomes you want which satisfies rule 3; every outcome has a 50% chance of occurring which satisfies rule 4.

Another example of ***what is <u>not</u> a binomial setting*** is being dealt three **red** cards in a row. This example fails rule 3 because after the first card has been dealt, the second card's probability of being red has changed from 26/52 to 26/51, and the third card's probability of being red would be 26/50. The sample space S changes when an event is removed after an outcome and that ***is not*** a binomial setting.

Binomial distributions

A ***trial*** is a "test" that examines outcomes. Let's pretend to perform a trial where we flipped a nickel 100 times and we counted 52 tails that came up. The number of successes (52 tails) in this trial is called the count. Now let's pretend that we did that same trial 23 times. Would we expect to get the exact same amount of successes (52 tails) in each and every one of the 23 trials? Nah! No way José! It is not impossible, but really close to it!

This different count from the different trials is called the ***random variable X***. Example: the third trial had 58 tails while the nineteenth trial had only 43. Since the chances of getting a head or a tail is 50%, we would naturally expect most of the trials to come up with exactly 50 tails, but this is not the case. There will be a few trials that will have a count of exactly 50, but most trials will not. Interesting and true!

This count of the different number of successes fits a binomial setting and therefore follows a ***binomial distribution***.

Binomial distributions have two parameters: ***n*** and ***p***. The number of individual trials is the ***sample size n***, and the ***probability*** of any single trial of being a success is ***p***. The specific number of successes, the random variable X, range from 0 to ***n***.

Once we know that the data fits a binomial distribution we can use the binomial probability formulas to predict long-term outcomes.

Binomial coefficients

Once we know that the data fits a binomial distribution we can now use the ***binomial coefficient*** as a first step to finding probabilities. The ***combination formula*** from the mathematical science of combinatorics tells us the number of how many different possible combinations there are of ***k success from n trials***. This is the number of all possible outcomes (the SRS) from the trials. The way we talk about this formula is ***n Choose k***, with **n** being all of the possible outcomes and **k** is the desired amount of outcomes.

Remember that a coefficient is simply a number multiplying a variable.

Here is the formula:

$$\binom{n}{k} = \frac{n!}{k!(n-k)!}$$

(k is a whole number ≤ n)

(n is the total number of trials)

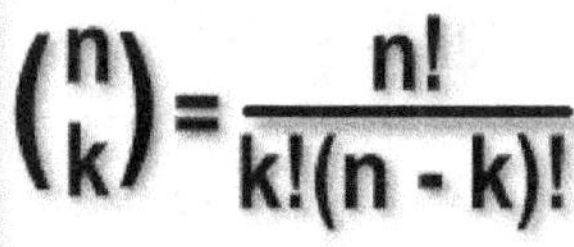

$$\binom{n}{k} = \frac{n!}{k!(n-k)!}$$

n is the total number of events and *k* is the specific number of desired outcomes.

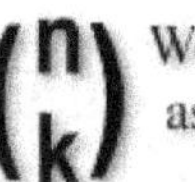

We pronounce this as ***"n choose k"***.

This can also be written as ***nCk***.

*Note: The exclamation mark means ***factorial*** which is a term from algebra that means to multiply repeatedly: **n! = n(n - 1)(n - 2)(n - 3)...(1).** Example: 5! = 5 x 4 x 3 x 2 x 1 = 120.

**Note: 0! = 1... just in case you were curious. Well, were you?

Online video example 6.1: One-eyed Pete's Pizzeria is having a sale on three topping pizzas. If there are ten toppings to choose from, how many different three topping pizzas can you choose from?

Step 1) Substitute the values for n and k: **n =10 and k = 3**

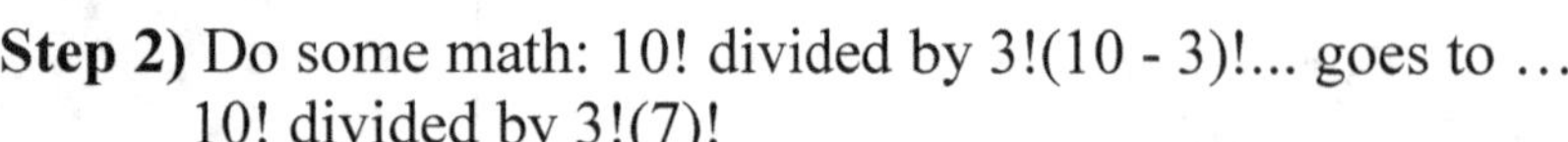

Step 2) Do some math: 10! divided by 3!(10 - 3)!... goes to …
10! divided by 3!(7)!

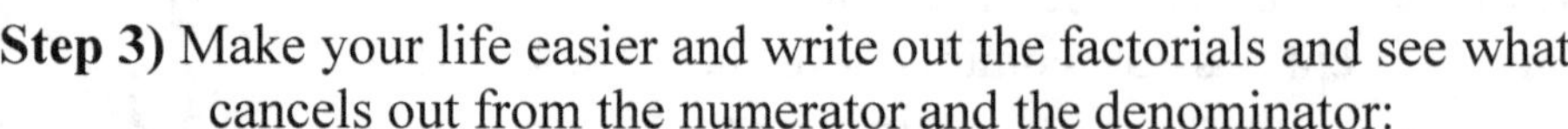

Step 3) Make your life easier and write out the factorials and see what cancels out from the numerator and the denominator:

(10 x 9 x 8 x 7 x 6 x 5 x 4 x 3 x 2 x 1)/(3 x 2 x 1)(7 x 6 x 5 x 4 x 3 x 2 x 1)

Step 4) Cancel out all of the factors possible:

(10 x 9 x 8 x ~~7 x 6 x 5 x 4 x 3 x 2 x 1~~)/(3 x 2 x 1)(~~7 x 6 x 5 x 4 x 3 x 2 x 1~~)
= (10 x 9 x 8)/(3 x 2 x 1)

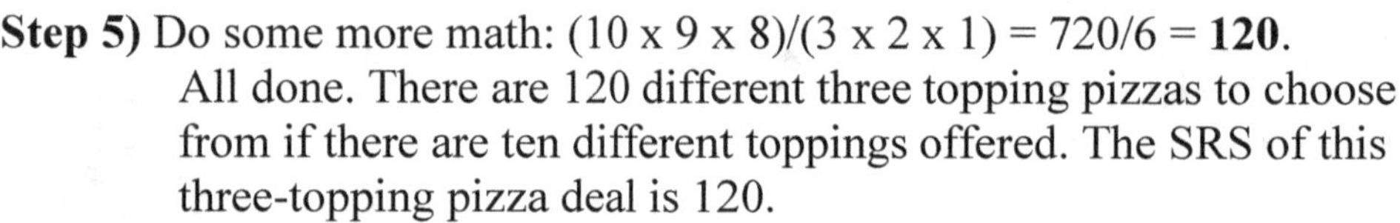

Step 5) Do some more math: (10 x 9 x 8)/(3 x 2 x 1) = 720/6 = **120**.
All done. There are 120 different three topping pizzas to choose from if there are ten different toppings offered. The SRS of this three-topping pizza deal is 120.

Note: This formula finds the total number of combinations ***regardless of order. In other words, a pineapple, pepperoni and onion pizza is considered to be the same as an onion, pineapple and pepperoni pizza. ***When order <u>does</u> matter***, then we use a slightly different calculation called a ***permutation formula***. Here is an example of a permutation problem: if ten people were in a foot race, what are the possible outcomes for who came in first, second and third place? In this case, Runner C (1st), D (2nd) and G (3rd) would be a different outcome than Runner G (1st), C (2nd) and D (3rd).

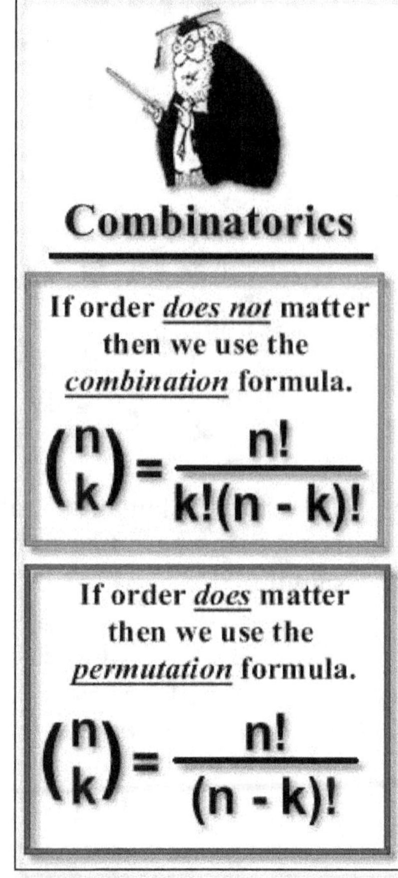

Binomial probabilities

The binomial coefficient is the first part of the binomial probability formula; it tells us the number of all the possible outcomes. The other part is the ***probability*** (***p***) of the desired outcome occurring ***k*** times out of the total number (***n***) of events. This is found by taking the probability of success for a single outcome (***p***) and raising it to the power that is equal to the number of desired successes (***k***): p^k. This number is then multiplied by the probability of any single outcome **<u>not</u>** being a success ($1 - p$) raised to the power of the difference between the total number of trials and the number of desired outcomes ($n - k$): $(1 - p)^{n-k}$. Confusing enough for you?

Here is the entire formula:

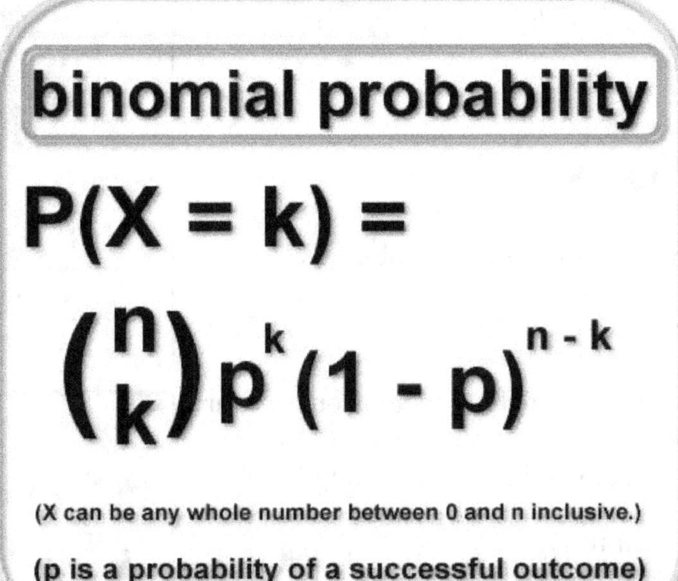

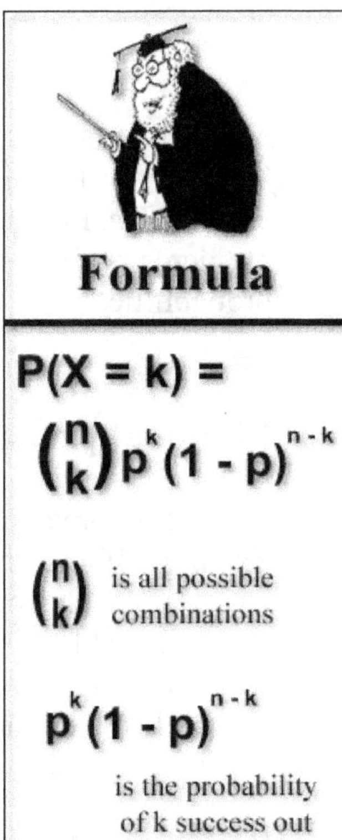

Online video example 6.2: Alex Rodriguez, the pride of the Yankees and the highest paid athlete of all time, has a lifetime batting average of .301. What is the probability that he will get exactly 2 hits out of his next 4 times at bat?

Step 1) Substitute the known values in the binomial probability formula:
P(X = 2) = (4 Choose 2)$(.301)^2(1 - .301)^{4-2}$

Step 2) Do some math, the binomial coefficient: (4 Choose 2) = 4!/2!(4 - 2)! = (4 x 3 x 2 x 1)/(2 x 1)(2 x 1) = 24/4 = **6**

Step 3) Do some more math, find the individual probability:
$(.301)^2(1 - .301)^{4-2} = (.301)^2(.699)^2 \approx$ **.0443**

Step 4) Do some more math: 6 x .0443 = .2658. So there is a **26.58%** probability that A-Rod will get exactly 2 hits out of 4 at bats.

*Important note: this formula finds the probability of EXACT counts only. In other words, the probability generated by this formula is ***not cumulative***. In order to find probability associated with AT LEAST problems we have to repeat the formula calculations and add the probabilities.

Binomial probabilities of AT LEAST problems

What if we are asked for the "***at least***" or "***at most***" probabilities? Example: If a baseball player has a lifetime batting average of .250, what are the chances that he will get ***at least*** one hit out of the next four attempts? At least one hit out of four at bats would be one hit, two hits, three hits and/or four hits. These are all acceptable outcomes out of four events. But the binomial probability formula only deals with exacts. This is where the math part of your brain should be kicking in… anything?

Yes! Good thinking. Just find the probabilities, using the binomial formula, for each and every set of outcomes that fit the desired outcome.

This is better explained with a worked example, so here we go.

Online video example 6.3: Archie the Archer has a lifetime record of hitting the target one time out of three shots. What is the probability that Archie will hit the target ***at least once*** out of three shots?

Step 1) Write out the values for the variables of the formula: probability for a successful outcome is one out of three (lifetime record), so **p = 1/3.**

The number of events is three shots, so **n = 3.**

The number of successful outcomes that we are looking for is at least 1, so we need to find, and add them up, the probabilities for Archie hitting the target once, twice and thrice (yes – thrice! It means three times but we don't use that word much nowadays):
k = 1 … k = 2 … k = 3, and then find the ***cumulative*** sum of **P(X = 1) + P(X = 2) + P(X = 3).**

Step 2) Figure out the probabilities for Archie hitting the target one time (once = 1). Substitute the known values in the binomial

If it is an *AT LEAST* probability: $P(X \geq k)$

you have to do this...

$$P(X = k) = \binom{n}{k} p^k (1 - p)^{n-k}$$

(probability of k successes)

+

$$P(X = k+1) = \binom{n}{k+1} p^k (1 - p)^{n-k}$$

(probability of k + 1 successes)

+

(... repeat this process all the way up to *n*)

$$P(X = n) = \binom{n}{n} p^k (1 - p)^{n-k}$$

(probability of n success)

This sum represents the cumulative probabilities of an AT LEAST **$P(X \geq k)$** problem.

probability formula beginning with 1:
$P(X = 1) = (3 \text{ Choose } 1)(1/3)^1(1 - 1/3)^{3-1}$ … goes to …
$(3)(1/3)(4/9) = 4/9$

Step 3) Figure out the probabilities for Archie hitting the target twice (twice = 2). Substitute the known values in the binomial probability formula beginning with 2:
$P(X = 2) = (3 \text{ Choose } 2)(1/3)^2(1 - 1/3)^{3-2}$ … goes to …
$(3)(1/9)(2/3) = 2/9$

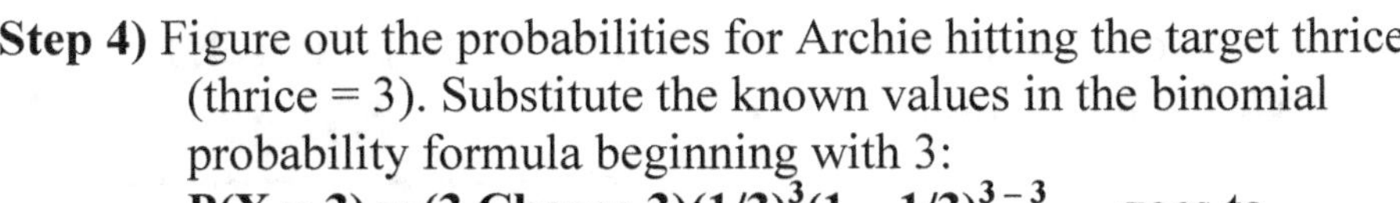

Step 4) Figure out the probabilities for Archie hitting the target thrice (thrice = 3). Substitute the known values in the binomial probability formula beginning with 3:
$P(X = 3) = (3 \text{ Choose } 3)(1/3)^3(1 - 1/3)^{3-3}$ … goes to …
$(1)(1/27)(1) = 1/27$

Step 5) Add up all the probabilities: 4/9 + 2/9 + 1/27 = **19/27**. This means that there would be around a 70% chance that Archie will hit the target at least once out of three shots.

Step 6) There is another way to solve these types of problems. Watch and learn grasshopper.

Remember this rule? For any event A, P(A) + P(not A) = 1.

Since the probability of Archie ***not*** hitting the target on any single shot is 2/3, then we could do this:

The probability of Archie not hitting any of the targets in three shots is the same as: P(not hitting target on the first shot) ***and*** P(not hitting target on the second shot) ***and*** P(not hitting target on the third shot).

Remember that in probability ***"and"*** means to multiply… P(not hitting target on the first shot) ***multiplied by*** P(not hitting target on the second shot) ***multiplied by*** P(not hitting target on the third shot)… goes to … (2/3) x (2/3) x (2/3) = **8/27**.

Since we know the odds of Archie not hitting the target in three shots, then the odds of Archie hitting it at least once is **1 – 8/27 = 19/27**. Got it?

*Note: for "at most" probabilities simply find the P(X = 0), P(X = 1), … P(X = k), add them all up and that will equal the cumulative probability.

Binomial mean and standard deviation

Once you are sure that the data fits the binomial distribution probability model, then the ***mean*** and the ***standard deviation*** are much easier to find. The mean is the total number of events (***n***) ***multiplied by*** the probability (***p***) of a successful outcome from any one event. For these types of problems, the mean is the average of many different trials and is the ***expected*** outcome for any trial.

Here are the formulas:

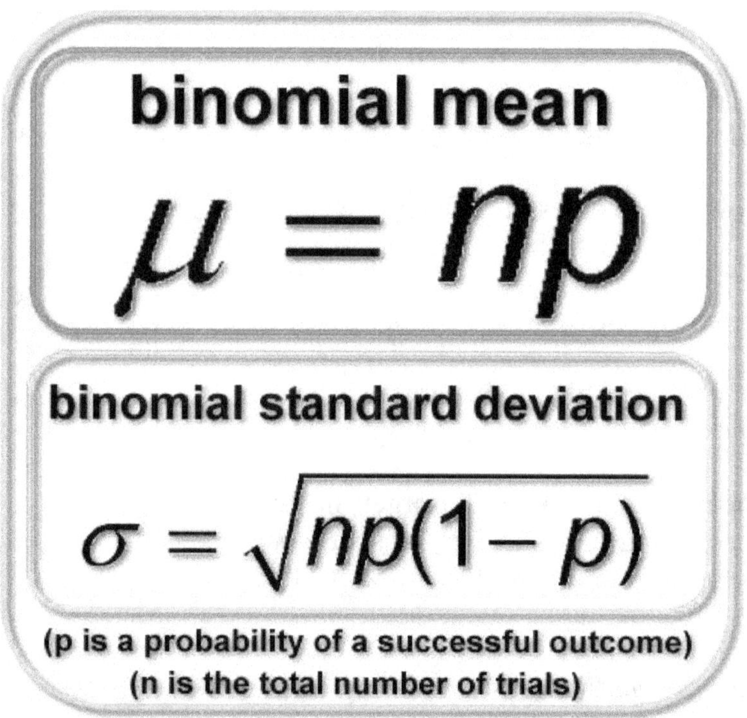

Binomial Mean

the number of events

$\mu = np$

the probability of a success from a single event

Online video example 6.4: Kobe Bryant of the LA Lakers has a career average of .840 of making his free throws. If he shoots 200 free throws in a season, how many (the ***binomial mean***) would you expect to go in?

Step 1) Since this is a binomial distribution, use the binomial mean formula: $\boldsymbol{\mu = np}$

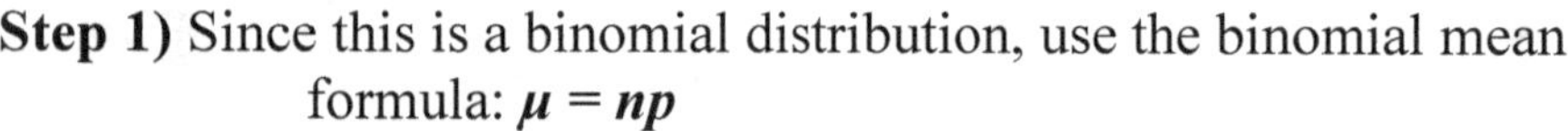

Step 2) Substitute known values and do the math: **μ = (200)(.840) = 168**

Step 3) Double check your answer: If he normally gets 84% of his free throws, is 84% of 200 equal to 168? Yeppers! This means that we would expect that Mr. Kobe would sink 84% of his free throws next year as well. But it is not guaranteed!

Online video example 6.5: Kobe Bryant of the LA Lakers has a career average of .840 of making his free throws. If he shoots 200 free throws in a season, what would you expect the ***binomial standard deviation*** to be?

Step 1) Since this is a binomial distribution, use the binomial standard deviation formula: **$\sigma = \sqrt{(np(1 - p))}$**

Step 2) Substitute known values and do the math:
$\sigma = \sqrt{(200 \times .840)(1 - .840))}$ = **5.185**

Chapter 6 Review

A ***binomial setting*** needs to meet these conditions: a large but finite sample size; no single outcome can influence any other outcome; each outcome has to be a "success" or a failure"; the probability of a successful outcome must be the same for each and every event.

A ***trial*** is a "test" that examines outcomes. The number of successes from a single trial is called the ***count***. The count from many trials is called the ***random variable X***, because the number of successes will be different from trial to trial.

This count of the different number of successes fits a binomial setting and therefore follows a ***binomial distribution***.

Binomial distributions have two parameters: ***n*** and ***p***. The number of individual trials is the ***sample size n***, and the ***probability*** of any single trial of being a success is ***p***. The number of successes, the random variable X, range from 0 to ***n***.

The ***binomial coefficient*** is calculated by finding all possible combinations of the outcomes (${}_{n}C_{k}$) and is the first part of finding the binomial probability. The other half of the formula is: $p^{k}(1 - p)^{n-k}$.

The entire binomial formula is $\mathbf{P(X = k) = ({}_{n}C_{k})(}\, p^{k}(1 - p)^{n-k}\mathbf{)}$

With a binomial distribution probability model, the ***mean*** and the ***standard deviation*** are much easier to find. The mean is the expected number of successful outcomes. The mean formula is $\mu = np$ with n being the number of events and p is the probability of a successful outcome from any one event. The standard deviation of a binomial distribution is $\sigma = \sqrt{(np(1 - p))}$.

mathguyzero

watch the online video explanations

Guided homework Chapter 6: Binomial Probabilities

1) Which of the following represents a binomial setting?

a) Taking 10 (playing) cards at random and assigning a "success" to a red card, then finding the probability that 7 out of ten of the cards would be red.

b) Blue eyes are the result of a gene from a parent. Geneticists tell us that a specific type of parent (let's say Italian) have the probability of 0.15 of having the "blue eye" gene. If these parents have 7 children, then the number of their 7 children that are born with blue eyes is the count X with the probability being 0.15 and the number of observations is 7.

c) Only 35% of telemarketing phone calls reach a live person. The telemarketing phone-calling machine makes 1,000 calls and X is the number of live people that answer.

2) If order does ***not*** matter, how many different three letter combinations can you make out of the following letters? A, B, C, D, E, F, G, H, I and J.

3) A recent study showed that the probability of randomly chosen homeowners has the probability of 0.80 that the value of their house has gone down since the previous year. Let X be the number of homeowners that have lost equity in their homes out of 25 randomly chosen homeowners.

a) Is this a binomial probability condition?

b) What is X (in words)?

c) What are n and p?

d) What are all of the possible values (SRS) of X?

e) What is the probability that X = 17?

f) What are the mean and standard deviation?

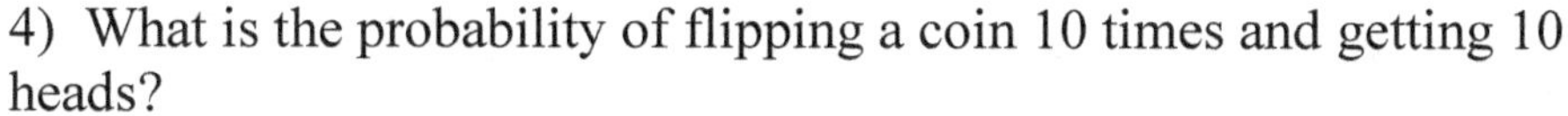
4) What is the probability of flipping a coin 10 times and getting 10 heads?

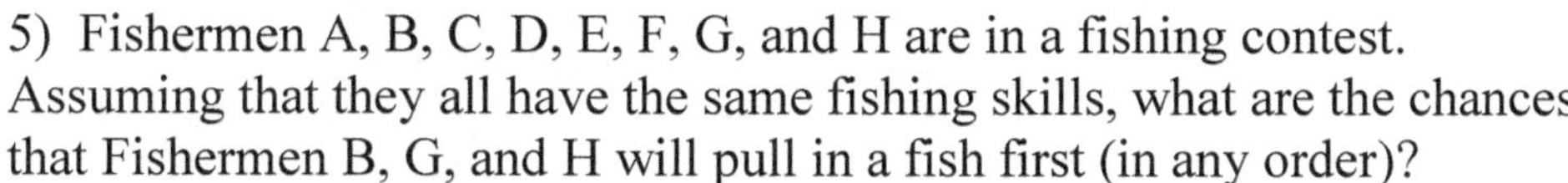
5) Fishermen A, B, C, D, E, F, G, and H are in a fishing contest. Assuming that they all have the same fishing skills, what are the chances that Fishermen B, G, and H will pull in a fish first (in any order)?

6) Manny Mota has a lifetime batting average of .300. What is the probability that he will get three hits out of five times at bat?

7) If a basketball player has a lifetime free throw record of 92%, how many would you expect him to make if he shot 85 free throws?

8) California has a lotto game called Fantasy Five where players pick 5 numbers from a field of 39 numbers. It is advertised as the game with better odds of winning. What are the chances of winning with a single ticket?

Chapter 6 Quiz

1. Which of the following is not a binomial setting?

    a. The probability of getting 7 heads out of 10 coin tosses
    b. The probability of getting three clubs in a row from a 52 card deck
    c. The probability of getting an odd number on a roulette wheel
    d. The probability of picking 5 numbers from a field of 39 numbers
    e. All of these are binomial settings

2. True or false?

The mean of a binomial setting is the sample size multiplied by the probability of a successful outcome from a single event.

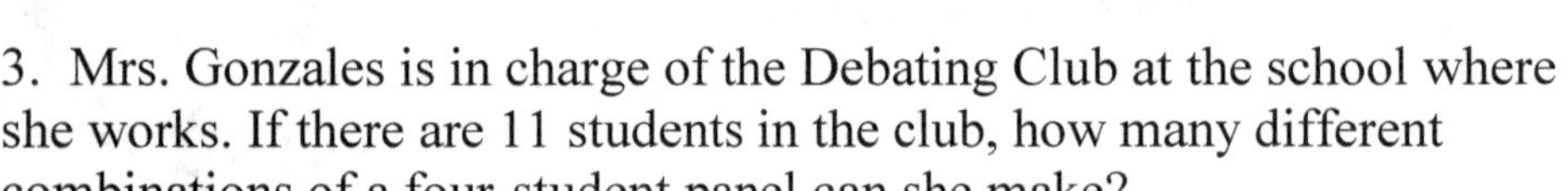

3. Mrs. Gonzales is in charge of the Debating Club at the school where she works. If there are 11 students in the club, how many different combinations of a four-student panel can she make?

4. A recent study showed that the probability of randomly chosen homeowners have the probability of 0.80 that the value of their house had gone down in value over the previous year.

    a. If you were going to call 25 of these homeowners at random to find out how many of them lost value in their homes, what exactly (in words) is the random variable X?

    b. If you were going to call 25 of these homeowners at random to find out how many of them lost value in their homes, how many different possible outcomes are there?

    c. If you were going to call 25 of these homeowners at random to find out how many of them lost value in their homes, how many of them would you expect to report a loss in value? (Hint: what is the mean?)

d. If you were going to call 25 of these homeowners at random to find out how many of them lost value in their homes, what would the standard deviation be?

e. If you were going to call 25 of these homeowners at random to find out how many of them lost value in their homes, what is the probability that you will get exactly 13 homeowners that lost value?

5. California has a Pick Three lottery game where the gambler picks three numbers from 0 - 9. What are the odds that a single ticket will win? (Hint: there are 10 numbers to choose from.)

6. Retired San Diego Padre Tony Gwinn Sr. has a lifetime batting average of .338. WOWZERS!!! That means that he got a hit close to once out of every three at bats. DOUBLE WOWZERS!!! That is impressive indeed. So, what were the chances that he would ***not*** get a hit during an entire game? Let's pretend that it was a 0 to 0 ballgame and he had 9 at bats, okee dokee... artichokee?

7. Gohkan says he is a psychic pizza maker. He claims that he can predict a customer's pizza order before they order a 5-topping pizza. What are the odds that he will get the order correct if the pizza restaurant offers 20 different toppings?

8. Fill in the blank:

The binomial probability formula uses the binomial coefficient and the  of the desired outcome.

a. individual event probability
b. binomial
c. setting
d. wanting
e. factor

9. Which of the following definitions fits best for "random variable X"?

a. The count of successful outcomes
b. The mean of a binomial setting
c. The binomial setting
d. The sample size multiplied by the probability of a binomial setting
e. The probability of a successful outcome

10. Back in WWII, British bomber pilots had a survival rate of 95% for every mission that they flew. What are the chances that a single pilot would survive long enough to fly 30 missions?

11. True or false?

In order to use the binomial probability formulas, there must be only two possible outcomes for any event or trial.

Ch 7: Sampling Distributions

Chapter 7

Sampling Distributions

Farmer Ted grows and sells potatoes. He wants to find the average weight of his potatoes. If he weighed just one potato would he get a mean weight of all of his potatoes? Probably not. What if he weighed 10 potatoes and divided that sum by 10? That **mean** would be closer to the **true mean** of all potatoes. What if he found the **mean** from 10,000 potatoes?

Finding the mean from a sample size that is greater than one is called ***sampling***. ***Sampling distributions*** work just like the standardized (**z**) scores from chapter 3 (normal distributions), but instead of comparing a single individual measurement to the mean of a group of measurements, we *compare the mean of a sample to the population's mean.*

In this chapter, we will investigate sampling distributions:

- Statistics and parameters

- Law of Large Numbers

- Sampling distributions

- Central limit theorem

- Sampling distribution probabilities

- Statistics and parameters
- Law of Large Numbers
- Sampling distributions
- Central Limit Theorem
- Sampling distribution probabilities

Statistics and parameters

Companies that advertise on television pay higher fees for their commercials on programs with higher Nielsen ratings. When the Nielsen Company states that 4.6 million households were watching a specific show on Tuesday night, do you think that they contacted all of the approximately 110 million households in the US and asked each of them what they were watching on their TVs? Nah! That is more crazy talk!

The Nielsen Company took a very large sample of households and made an ***inference*** about the entire population. The descriptive data from the sample are called ***statistics***, also known as ***descriptive statistics***, but the descriptive information regarding entire ***populations*** are called ***parameters***. Parameters are represented with Greek letters and statistics are represented with English letters.

*Note: Parameters are rock-solid facts, but statistics are used to describe and predict, and that is why you are reading this sentence right now, learning how this predicting is done... with numbers! Feeling the math?

The science of statistics deals with making predictions about the population from a sample, or many samples, of that population. This art of predicting what is true for the population from a sample is called ***inferential statistics***. This is the science behind insurance.

Law of Large Numbers

Back to the Nielsen TV ratings; imagine how accurate their ratings would be if they could contact all 110 million households. The data that they had collected would be parameters and there would be no doubt of how many people would be watching which shows. Soon, this option will be a possibility with emerging technologies, especially when the television gets absorbed by the computer... predicts mathguyzero.

The population mean and standard deviation exist. If we were able to weigh every apple on earth, then we would have the population mean and standard deviation. But, if we picked just one apple at random, would it be a good representation of the average apple? Maybe, but what if it just happened to be the biggest apple in the world?... or the smallest?

If we picked 10 apples at random, their *average weight* and *standard deviation* would be closer to the true population mean and standard deviation, because 10 is a larger part of the population than the single apple. If we picked 10,000 apples at random, the statistics from this sample would be closer still to the true population mean and standard deviation, again because the sample size is closer to the population size.

When it comes to statistics, ***the larger the sample size the better***. We call this rule the ***Law of Large Numbers***. Think about it this way: if you had a sample size as big as the entire population, then your mean and standard deviation would no longer be statistics, but parameters, and that is what we are searching for using the sampling method.

Law of Large Numbers

The Law of Large Numbers basically states that the larger the sample size the more accurate the statistics will be compared to the population parameters.

In other words, the bigger the sample size, the better.

Sampling distributions

Back to the example of all the apples in the world; if you took a random sample of 10 apples from California and another random sample of 10 apples from Washington, would you expect the means and standard deviations to be the same from the two samples? Nah, not very likely.

In fact, you would get a little bit of natural (normal) variance in every sample of 10 apples that you took. This taking of a sample over here, a sample over there and another sample from way over there, where all of the samples sizes are the same, is called ***sampling***.

These unequal means and standard deviations from the different samples are called ***sampling distributions*** and will fit a normal distribution. Therefore, we can use the same rules from the standardized normal distribution formulas, but instead of a *random variable x* we use the **sample mean**, and we change the standard deviation to the standard deviation divided by the square root of the sample size (**standard error**). This rule is called the **Central Limit Theorem.**

Central Limit Theorem

Central Limit Theorem

If a random sample $\bar{x}$ of size n is taken from a population with a mean of μ and a standard deviation of σ, then the sampling distribution of $\bar{x}$ has a mean of μ and a standard deviation of $\frac{\sigma}{\sqrt{n}}$.

Example: What would the mean and standard deviation from a sample size of 50 be if the population mean was 100 and the standard deviation of 15, N(100, 15)? (A standard deviation divided by the square root of the sample size is called a **standard error**.)

Answer: Because the sample size is large, the sample mean used would be the population mean, and the standard deviation would be the population's standard deviation divided by the square root of the sample size. N(100, 15/√50) = **N(100, 2.12)**

Sampling distribution probabilities

The unequal means from the many different samples of the same size (the ***sampling distribution***) will fit a normal distribution curve. We can standardize the sample means with the standardization formula, but with a few changes. We replace the ***random variable x*** with the ***sample mean***, and replace the standard deviation with the standard deviation divided by the square root of the sample size (**standard error**). And since the data fits a normal distribution curve, we can use the normal distribution probability formula.

Here is the formula:

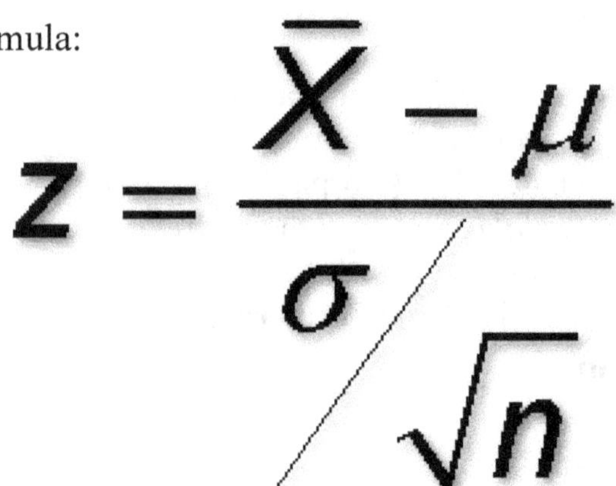

$$z = \frac{\bar{X} - \mu}{\sigma / \sqrt{n}}$$

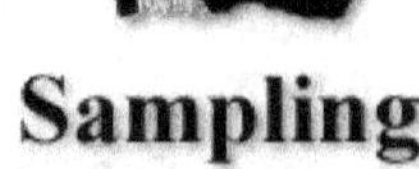

Sampling compares a group mean to a population mean instead of comparing an individual measurement to a group mean.

We still use the properties and formulas of a normal distribution from chapter 3 with two exceptions: substitute the random variable **x** with the sample mean $\bar{X}$, and the σ with $\sigma/\sqrt{n}$.

Online video example 7.1: Farmer John has been growing avocados for 38 years. He has recorded the weights every year and knows that they the fit a normal distribution curve with a mean weight of 7.6 ounces with a standard deviation of 1.2 ounces. He sends his workers to bring back many boxes of avocados, 25 per box that were picked at random, from all over his 600 acre ranch. He finds the avocados' mean weight from each of the boxes of 25 avocados.

How many of the sample means will be over 7.6 ounces?

Answer 1) Trick question: because if the weights fit a normal distribution curve, then we already know that **50%** of the sample means will be greater than the mean (7.6 oz) and 50% will be less than the mean. But you knew that already, ja?

Online video example 7.2: What is the probability that a sample mean from one of the boxes will be less than 7.0 ounces?

Answer 2) Standardize the data and look up the z score in the z table or on Excel: $\mathbf{z = (7.0 - 7.6)/ ((1.2)/(\sqrt{(25)}) = -2.5}$ which generates an area of **.0062** under the probability curve. So, there is a **.62%** percent chance that a sample of 25 avocados will have a mean weight of 7.0 oz or less. Not very likely, eh?

Online video example 7.3: What is the probability that a sample mean from one of the boxes will be greater than 7.9 ounces?

Answer 3) Standardize the data and look up the z score in the z table or on Excel: $\mathbf{z = (7.9 - 7.6)/ ((1.2)/(\sqrt{(25)}) = 1.25}$ which generates an area of **.8944** under the probability curve. But we need to **subtract this number from 1 because it is a greater than problem**. So, 1 - .8944 = **.1056** and that tells us that there is a **10.56%** percent chance that a sample of 25 avocados will have a mean weight of 7.9 oz or greater.

Online video example 7.4: What is the probability that a sample mean from one of the boxes will be between 7.2 oz and 7.5 oz?

Answer 4) This problem will generate a range of values because it is asking for the probability between two weights. We start with the largest value first: $\mathbf{z = (7.5 - 7.6)/ ((1.2)/(\sqrt{(25)}) = -.42}$ which generates an area of .3372 under the probability curve. Next we repeat the process with the smaller value: $\mathbf{z = (7.2 - 7.6)/ ((1.2)/(\sqrt{(25)}) = -1.67}$ which generates an area of .0475 under the probability curve. Now we **subtract** the smaller value from the larger value and that will give us the area under the curve that corresponds to the problem.

So, .3372 - .0475 = **.2897** and that tells us that there is a **28.97%** percent chance that a sample of 25 avocados will have a mean weight between 7.2 and 7.5 ounces.

Chapter 7 Review

Chapter Review

Comparing the mean from a sample size that is greater than one to a population is called ***sampling***. Sampling distributions work just like the standardized (z) scores and fit a normal distribution, but instead of comparing a single individual measurement to the mean of a group of measurements, we compare the mean of a group (sample) to the population's mean.

Information derived from a sample is called a ***statistic*** and information from a population is called a ***parameter***. Parameters are represented with Greek letters and statistics are represented with English letters.

Descriptive statistics describe large amounts of data. ***Inferential statistics*** makes predictions about an entire population using information (statistics) from a sample.

When it comes to statistics, ***the larger the sample size the better***. We call this rule the ***Law of Large Numbers***.

The ***Central Limit Theorem*** states that if you have a sample of sufficient size ***n*** from a population, then you can assume that the sample mean is equal to the population mean, and that the sample standard deviation is equal to the population's standard deviation divided by the square root of the sample size (**standard error**).

Sampling probabilities will fit a normal distribution curve and therefore we can use the same formulas and tables from chapter 3. We can standardize the sample means with the standardization formula, but with a few changes. We replace the ***random variable x*** with the ***sample mean***, and replace the standard deviation with the standard deviation divided by the square root of the sample size.

Sampling

Sampling compares a group mean to a population mean instead of comparing an individual measurement to a group mean.

We still use the properties and formulas of a normal distribution from chapter 3 with two exceptions: substitute the random variable **x** with the sample mean $\bar{x}$, and the σ with $\sigma/\sqrt{n}$.

Guided homework Chapter 7: Sampling Distributions

A farmer harvests his apple crop and weighs them in grams. Their weights fit a normal distribution of N(52, 14). What is the probability that:

1. One apple selected at random will weigh over 60 grams?

2. Four apples will have an average weight of less than 45 grams?

3. One hundred apples will have a mean weight between 50 and 52 grams?

4. One thousand apples will have a mean weight of more than 52.1 grams?

The stalks of sugar cane in a sugar cane plantation have a mean height of **5.3** feet with a standard deviation of **1.5** feet. Assume the heights are normally distributed. We randomly select the stalks **fifty** at a time and measure them.

5. If we sampled **all** of the sugar cane stalks 50 at a time, what would the sample mean be?

6. What is the standard error of the sample means?

7. What is the probability that a sample mean will be less than 5.5 feet?

8. What is the probability that a sample mean will be between 5.0 feet and 5.5 feet?

9. What is the probability that a sample mean will be greater than 6 feet?

Chapter 7 Quiz

1. A farmer harvests his orange crop and weighs them in grams. Their weights fit a normal distribution of N(144, 28). What is the probability that:

a. One orange selected at random will weigh over 160 grams?

b. Four oranges will have an average weight of less than 145 grams?

c. One hundred oranges will have a mean weight between 150 and 160 grams?

d. One thousand oranges will have a mean weight greater than 144.5 grams?

e. Ten thousand oranges will have a mean weight of less than 144 grams?

2. The rule that states that the bigger the sample size, the closer that the calculated statistics will be to the true population parameters, is called the

_________________ .

a. Probability Limit Theory
b. General Rule of Probability
c. Law of Large Numbers
d. Law of Sample Size
e. Parameters are Closer Than You Think

3. Intelligent quotas (IQs) are normally distributed with a mean of 100 and a standard deviation of 15. According to the Central Limit Theorem, if you took sample of 25 people at random, what would the standard error be?

4. The Central Limit Theorem states that when you take a sample from a population, go ahead and use the sample mean instead of the population mean in the normal distribution formulas, and then divide the population standard deviation by the square root of the ________ ________ .

a. sample size
b. average mean
c. stated norm
d. binomial probability
e. general probability

5. The statistical science of making predictions with long-term probabilities is called ________ statistics.

a. quantitative
b. inferential
c. explanatory
d. parametric
e. descriptive

6. True or false?

Statistics are derived from samples and are represented with letters from the English alphabet.

7. Intelligent quotas (IQs) are normally distributed with a mean of 100 and a standard deviation of 15. According to the Central Limit Theorem, if you took sample of 25 people at random, what should the sample mean be?

8. True or false?

Parameters are from populations and are represented with letters from the Greek alphabet.

Ch 8: Confidence Intervals

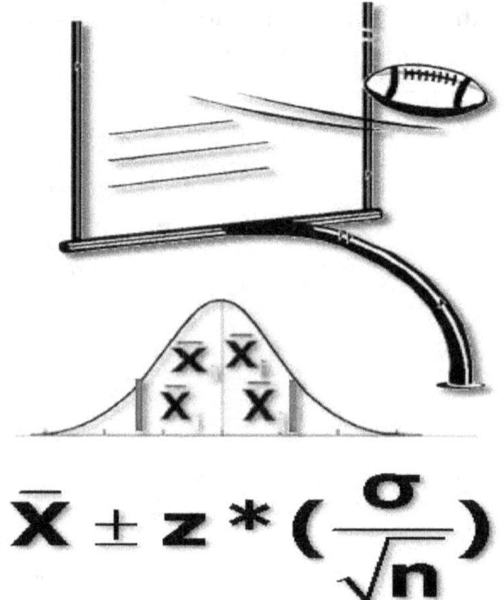

Chapter

8

Confidence Intervals

The 2010 California State STAR test scores reported that 9% of the 11th graders from the second largest school district in California scored at the level of proficient or above in Algebra I. The report states that 56,797 11th graders took the test. Since this sample was so large, we can invoke the Central Limit Theory and assume that the population mean of all 11th graders would be the same as this sample: 9% of all 11th graders will score at the level of proficient or higher. This assumption is called an ***inference***.

How accurate is a sample mean compared to the true population mean? If the same research were repeated a million times, how frequently would you expect to get similar results? How many of those trials would give some extreme result? With statistics from a sample we can calculate a range, or an interval, of values that should "**capture**" the true mean of the population. We call it a ***confidence interval***.

In this chapter we will investigate:

- Conditions for a confidence interval

- Confidence levels

- Margin of error

- Confidence intervals

- Sample size

- Conditions for a confidence interval
- Confidence levels
- Margin of error
- Confidence intervals
- Sample size

Conditions for a confidence interval

We construct a ***confidence interval*** (***CI***) to estimate, or ***infer***, the true mean of a population using statistics from a sample. We use confidence intervals when testing null hypotheses which we will cover in the next chapter.

In order to feel *confident* about our confidence intervals we need to set the ground rules first. Remember always that unless the following conditions are met, the data that is derived from these statistics formulas will not be valid. Which means that you have wasted your precious time… Bah!

Condition 1) The sample that we use must be chosen at random - a simple random space – SRS. (Always ***random*** – no bias.)

Condition 2) The random variable, or sampling mean, must have a normal distribution in the population - N(μ,σ). (Data has to be ***normal***.)

Condition 3) The true population mean must be **unknown** (μ), but we must know the population standard deviation (σ). (This is the weird rule.)

*Note: These three conditions must always be suspect. How random is random? How well does the data fit a normal distribution? And if we don't know the population mean, how are we supposed to know the population standard deviation? It takes a mean to derive a standard deviation, yes? Well, mgz is waiting…

Sample Random Space (SRS)

An SRS is a randomly chosen sample from a population.

An SRS can also be called a "simple" random space. The two definitions mean the same thing: no bias!

Confidence levels

The ***level of confidence*** depends on the researcher. Normally, we want to be very confident so we choose confidence levels of 90% or higher. After all, why would anyone want to be only 30% confident?

A 90% confidence interval translates into "*If we repeated this research study under the same conditions each time, then the mean from these repeated studies of samples should fall within this interval 90% of the time*", and therefore the true population mean must be within this range of values 90% of the time as well. That is the theory.

We label the different confidence levels as ***confidence level C***. The most popular confidence interval is 95%. ***<u>There is always a 5% probability that the true mean will not be in the 95% confidence interval.</u>***

Alpha

Alpha (α) is the area under that probability curve that falls outside of the confidence interval.

Remember that 100% - CI = α

This "left-over" probability is called ***alpha (α)***. Remember this always: the ***level of confidence C + alpha (α) = 100%.***

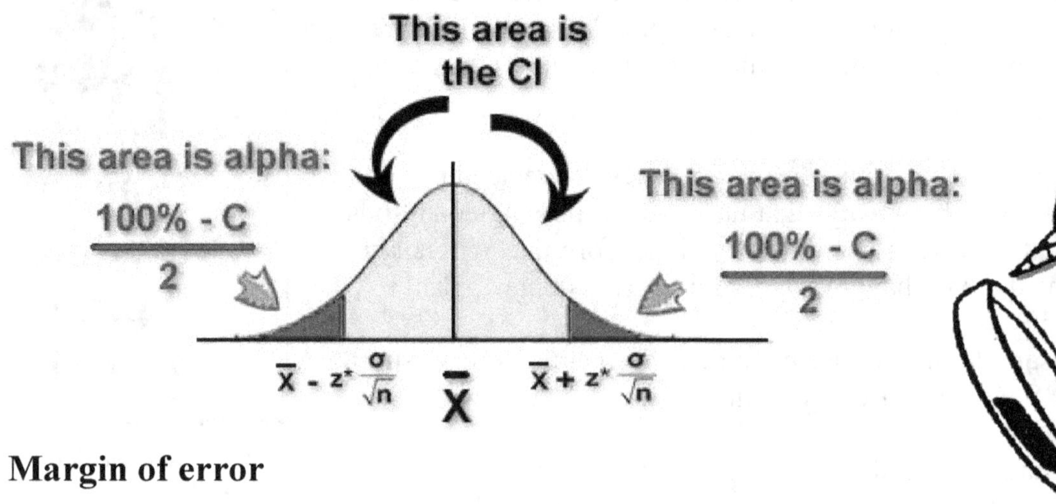

Margin of error

There are two parts to building a confidence interval. We start with the **sample mean** (remember that we are looking for the true population mean so we call the sample mean the "**estimate**" of the population mean) as the **center** of our confidence interval. We then add and subtract the ***margin of error*** to, and from, the sample mean. The greater end of the interval is the ***sum*** of the sample mean and the margin of error, while the smaller end of the interval is the ***difference*** between the sample mean and the margin of error. We will use the letter **m** to abbreviate the margin of error; other instructors use the letter E.

Alpha

Alpha (α) is the area under that probability curve that falls outside of the confidence interval.

Remember that 100% - CI = α

Confidence interval = x-bar ± margin of error

Finding the sample mean is basic, but finding the margin of error depends on the level of confidence C that we want to construct.

*Note: the abbreviation of a confidence interval is CI. The CI is very closely related to the level of confidence C, but they are not exactly the same thing. A confidence interval is a range of specific values and the level of confidence is kind of a control factor because the higher the level of confidence C, the larger the ***critical value (z*)*** used in the CI formula.

**Note: the greater the level of confidence C, the wider the range of the confidence interval.

Here is the formula to find the margin of error depending on the level of confidence.

margin of error

$$\text{margin of error} = \pm\, z^* \frac{\sigma}{\sqrt{n}}$$

z* is the critical value corresponding to the confidence level C

The level of confidence C tells us which critical value (**z***) from the table to use when we build our margin of error. (These values are from the normal distribution (z) table.)

confidence level C	90%	95%	99%
critical value z*	1.645	1.960	2.576

Online video example 8.1: Make a margin of error for a 95% level of confidence C with the $\sigma = 5$ and a sample size of 100.

Step 1) Break out the margin of error formula: $\mathbf{m = \pm z^* (\sigma/\sqrt{n})}$

Step 2) Substitute known values into the formula: $\mathbf{m = \pm 1.960 (5/\sqrt{100})}$
(Why did $z^* = 1.960$ in this example? Because 1.960 is the corresponding critical value (from z table) for the 95% CI.)

Step 3) Math time: $\mathbf{m = \pm 1.960 (5/10)}$ … goes to …
$\mathbf{m = \pm 1.960 (.5) = \pm 0.980}$

Online video example 8.2: Make a margin of error for a 99% level of confidence C with the $\sigma = 1.2$ and a sample size of 50.

Step 1) Break out the margin of error formula: $\mathbf{m = \pm z^* (\sigma/\sqrt{n})}$

Step 2) Substitute known values into the formula: $\mathbf{m = \pm 2.576 (1.2/\sqrt{50})}$
(Why did $z^* = 2.576$ in this example? Because 2.576 is the corresponding critical value (from z table) for the 99% CI.)

Step 3) Math time: $\mathbf{m = \pm 2.576 (1.2/\sqrt{50})}$ … goes to …
$\mathbf{m = \pm 2.576 (.1697) = \pm .4371}$

Online video example 8.3: Make a margin of error for a 90% level of confidence C with the $\sigma = .046$ and a sample size of 1,000.

Step 1) Break out the margin of error formula: $\mathbf{m = \pm z^* (\sigma/\sqrt{n})}$

Step 2) Substitute known values into the formula:
$\mathbf{m = \pm 1.645 (.046/\sqrt{1000}) = m}$
(Why did $z^* = 1.645$ in this example? Because 1.645 is the corresponding critical value for the 90% CI.)

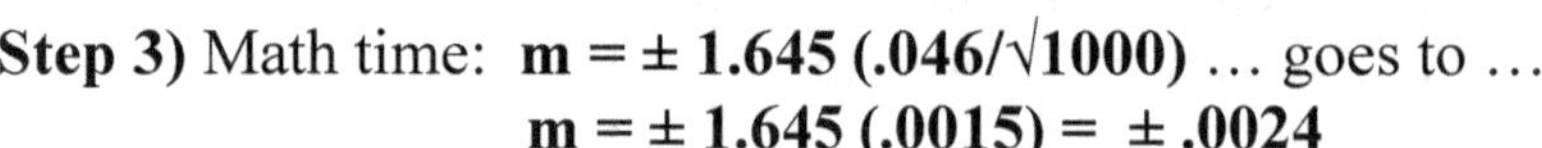

Step 3) Math time: $\mathbf{m = \pm 1.645 (.046/\sqrt{1000})}$ … goes to …
$\mathbf{m = \pm 1.645 (.0015) = \pm .0024}$

Confidence intervals

The information needed to teach a mathematical concept is embedded in the vocabulary that is used to describe and explain it. The vocabulary words of ***confidence interval*** (CI) are a perfect example. We will build an interval, or a range of values, with which we will be confident (pretty sure, not 100% though) that the true population mean will lie between.

Confidence intervals have two purposes and one of them is to try and "**trap**" the true population mean between two values. We use this fact in the next chapter to determine if a hypothesis is worthy of our belief… or not. We will use a formula to get a probability value from a collected sample and **if the p-value falls within the CI, then the stated population mean is probably true**, but if the p-value falls outside the CI, then the stated population mean is probably not true. It is a totally cool process!

The two parts needed to build a confidence interval are the sample mean and the margin of error.

Here is the formula:

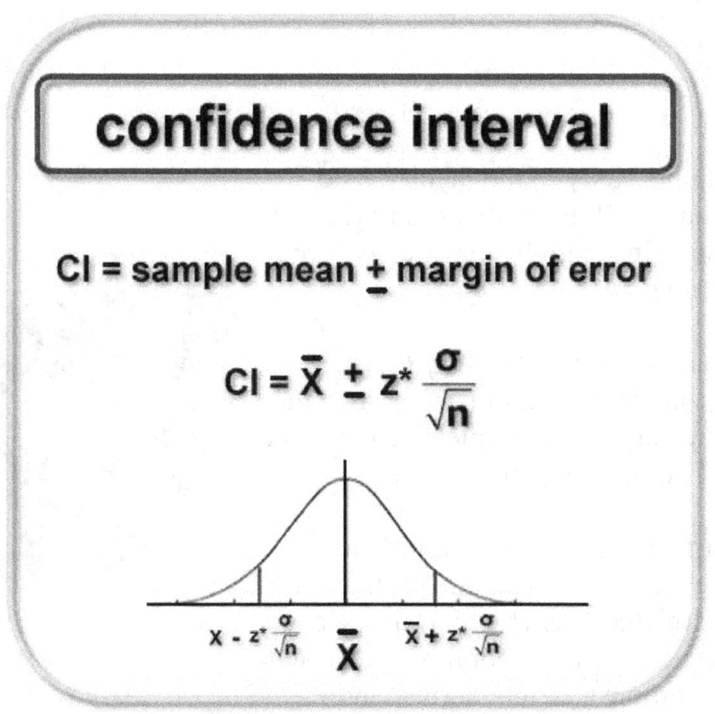

In the confidence interval formula for a z test, z* represents the critical value (from a table) corresponding to a a level of confidence.

*Note: The level of confidence C tells us which critical value (**z***) to use when we build our margin of error.

Online video example 8.4: Build a 95% confidence interval for a sample mean of 12.5, a population standard deviation of 2.8 with a sample size of 40.

Step 1) Break out the CI formula: the sample mean = **12.5**, but we will have to figure out the margin of error.

Step 2) Build the margin of error: z* = 1.960 (because that is the critical value that corresponds with a 95% CI), sigma (σ) = 2.8 and n = 40... so the margin of error = 1.960(2.8/6.325) = **.8677**.

Step 3) Add the margin of error to the sample mean for the greatest value and subtract it from the mean for the smallest value:
CI = [11.6323, 13.3677], or we can write it as **12.5 ± .8677**.

Step 4) Interpret the CI: *if we repeated this same study over and over again under the same conditions, then the sample means should fall between this range of values of [11.6323, 13.3677] 95% of the* *time. Based on this data, we presume that the true population mean is somewhere between 11.6323 and 13.3677.*

Sample size

Imagine a neophyte researcher presenting a study to the faculty. The researcher goes on and on about the study until the professors find out that all of the data that the researcher had so painstakingly analyzed was only obtained from two subjects. Oh, the shame!

The ***Law of Large Numbers*** tells us that the bigger the ***sample size*** the closer the sample statistics (i.e., mean, standard deviation) will be to the true population parameters. But just how large of a sample size is needed to ensure accuracy? What is the cut-off number for sample size?

Did you notice that the margin of error formula from the confidence interval has the sample size ***n*** in it? Well, did you? That means that the sample size and the margin of error are related to each other, like second cousins. From the margin of error formula we can algebraically rearrange it to give us the formula for the correct sample size to a corresponding margin of error. Ahhhh… the power of algebra!

Here is the sample size formula:

$$n = \left(\frac{z^*\sigma}{m}\right)^2$$

z^* is the critical value corresponding to the confidence level C

m is the $\pm$ of units that are being used in the specific margin of error

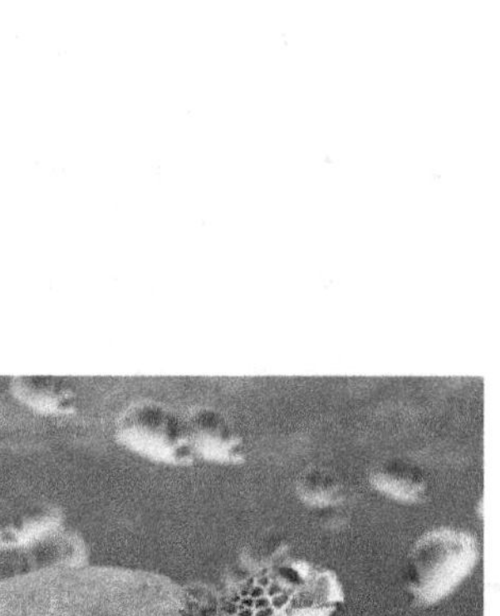

Example 8.5: A research biologist studying squirrels weighed many of the cute little furry guys and got a sample mean weight of 48.6 grams with a standard deviation of 4.8 grams. The scientist wants to make an estimate of the true population mean within a margin of error of only 0.1 grams with 95% confidence. How many squirrels should he weigh?

Step 1) Break out the sample size formula: $\mathbf{n = ((z^*\sigma)/m)^2}$

Step 2) Substitute known values: $\mathbf{n = ((z^*\sigma)/m)^2}$= $((1.960)(4.8)/0.1)^2$

Step 3) It's do the math time: $((1.960)(4.8)/0.1)^2 = (94.08)^2 = 8851.0464$

Step 4) When dealing with sample sizes, ALWAYS ROUND UP TO THE NEAREST WHOLE NUMBER!!! After all, .0464 of a squirrel sounds kind of messy, don't you think? So, the correct number of squirrels for this study is **8852**. That's going to take a lot of nuts!

Chapter 8 Review

A ***confidence interval*** is used to determine the true mean of an entire population using statistics from a sample of that population. A confidence interval is the estimate (the sample mean, **x-bar**) **± margin of error**. We use confidence intervals to reject (or not) ***hypotheses*** which we will cover in the next chapter.

Sample Random Space (SRS)

An SRS is a randomly chosen sample from a population.

An SRS can also be called a "simple" random space. The two definitions mean the same thing: no bias!

Conditions for a confidence interval (CI): 1) The sample that we use must be chosen at random - a simple random space – SRS; 2) The random variable, or sampling mean, must have a normal distribution in the population - $N(\mu,\sigma)$; and 3) The true population mean must be unknown (μ), but we must know the population standard deviation (σ).

The ***level of confidence C*** corresponds to how confident the researcher wants to be. The most used levels of confidence are 90%, 95% and/or 99%. Levels of confidence tell us which critical value (z*) to use.

The ***margin of error*** is the plus or minus amount that is added and subtracted from the sample mean. Margin of errors are calculated from a formula which depend on the level of confidence C.

The minimum ***sample size*** for a confidence interval is derived by a formula and depends on the level of confidence C and the size of the margin of error.

Guided homework Chapter 8: Confidence Intervals

1. Three hundred 8th graders were given an IQ test. The mean of their scores was 103.6 with a standard deviation of 4.8. Assume that the scores were normally distributed and give a confidence level of 99% for the true population IQ mean.

2. The weights (in grams) of Hershey's Kisses are normally distributed. A sample of 100 Kisses has a mean weight of 8.2 grams with a standard deviation of 1.2 grams.

a. What is the margin of error for a 90% confidence interval?

b. What is a 90% confidence interval for the mean weight of the Kisses?

c. What are you 90% confident about? (What happens 90% of the time?)

d. What is a 99% confidence interval for the mean weight of the Kisses?

3. A random sample of 32 nails have the following lengths in centimeters. Assume that the lengths are normally distributed. What is a 95% confidence interval for the mean length of all the nails?

7.65	7.62
7.87	7.55
7.04	7.50
7.41	7.48
8.02	7.77
7.66	8.05
7.60	8.00
7.91	7.32
7.01	7.90
7.85	7.44

6.98	7.51
7.15	7.45
7.13	7.68
7.73	7.81
7.34	7.92
7.44	7.22

4. You have been hired by the government to check the accuracy of gasoline pumps at local gas stations. You pump out exactly one gallon, according to the pump, into your scientific bucket, and then pour the gasoline into your super accurate measuring machine. You do this 50 separate times and the average amount of gasoline is .9743 gallons. You know that the pump has a standard deviation of 0.02 gallons.

a) Give a 90% confidence interval for the mean of every (supposed) gallon that this pump actually puts out.

b) How many measurements must be averaged to get a margin of error of ± .001 with 98% confidence?

5. Three hundred 8th graders were given an IQ test. The mean of their scores was 103.6 with a standard deviation of 4.8. What should the sample size be if you wanted to estimate the population mean within ± 0.5 points with 99% confidence?

Chapter 8 Quiz

1. A random sample of hens' eggs have the following weights in grams. Assume the weight of chicken eggs fits a normal distribution and build a 90% confidence interval for the true population mean for hen's eggs.

29.6	**28.4**
30.2	**29.6**
28.6	**30.9**
27.5	**26.1**
32	**27.1**
33.1	**30.8**
29.8	**29.1**
27.5	**28.4**
28.8	**31.6**
25.9	**30.2**
27.3	**32.7**
30.8	**26.5**
31.5	**27.6**
27.3	**29.4**
29.4	**27.6**
27.6	**30.1**
27.4	**31.9**

2. True or false?

When building a confidence interval, it is extremely important that the sample means fit a normal distribution curve.

3. Which of the following is the corresponding critical (z*) value for a 95% level of confidence C?

a. 1.645
b. 1.558
c. .95
d. 1.960
e. 2.576

4. What size confidence interval would you construct if you had a critical value alpha (α) of .05?

5. An ichthyologist weighed 100 four-foot squids (ewwww!) and got a sample mean of 26.35 lbs with a standard deviation of 3.87 lbs. He wants to estimate the true mean weight of four-foot squids with a margin of error of 0.2 lbs with a 95% confidence interval. How many squids should he weigh?

6. Build a 98% confidence interval from a sample of 1,000 with a sample mean of 89.67 and a population standard deviation of 7.35.

7. True or false?

If the critical value alpha (α) is not mentioned, we always use the value of .01.

8. Make a margin of error for a 95% confidence level with a standard deviation of 65 and a sample size of 1,000.

9. True or false?

In order to build a confidence interval, first you must know the population mean.

10. If you built a 99% confidence interval, then the critical value alpha (α) would be what?

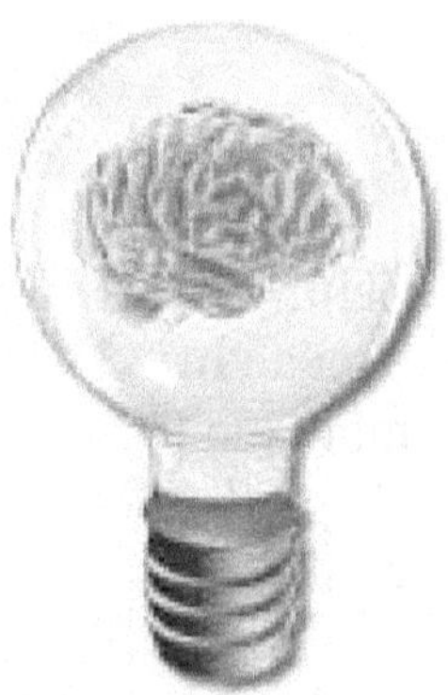

11. Construct a 95% confidence interval for a sample of data that has a sample mean of 55 grams with a known population standard deviation (σ) of 5 grams and a sample size of 40.

12. Which of the following is the corresponding level of confidence C for a critical (z*) value of 1.645?

a. 95%
b. 100%
c. 62.5%
d. 90%
e. 99%

13. Construct a 99% confidence interval from a sample size of 100 with a sample mean of 69 and a population standard deviation of 2.

14. What falls between the two values of a 95% confidence interval?

a. 95% of the different sample means
b. 95% of the data
c. 95% of the measurements will be less than the mean
d. 95% of the standard deviation
e. 95% of the margin of error

15. Make a margin of error for a 99% confidence level with a standard deviation of 7.2 and a sample size of 36.

16. True or false?

The critical value alpha (α) is equal to the difference between 100% and the confidence interval.

chapter quiz

17. True or false?

If the stated, or claimed, population mean is indeed true, then the sampling means will fall between the confidence interval values 90% of the time, 95% of the time, or whatever percent of the time that corresponds to the level of confidence C.

18. A zoologist weighed 10 baby brown bears and got a sample mean of 12.8 lbs with a standard deviation of 2.6 lbs. She wants to estimate the true mean weight of baby brown bears with a margin of error of 0.2 lbs with a 90% confidence interval. How many baby bears should she weigh?

19. A confidence interval is the sample mean plus/minus the ______.

a. margin of error
b. standard deviation
c. variance
d. level of confidence
e. critical value

Ch 9: Null Hypothesis – The Basics

Chapter

9

Null Hypothesis
- The Basics

We use confidence intervals for two types of statistical inferences: to estimate a true population mean, and for ***tests of significance***. Tests of significance are a way to look for evidence calculated from the sample data that supports, or does not support, a claim about a population parameter, usually the mean. The statement being tested is called the ***null hypothesis***.

In this chapter, we will investigate the validity of a hypothesis by calculating a confidence interval from a sample to see if there is a ***significant*** difference between the sample data and the stated population mean.

- Tests of significance

- Null hypothesis

- Test statistics and their corresponding p-values

- Significance

- Interpreting with a graph

- Tests of significance
- Null hypothesis
- Test statistics and their p-values
- Significance
- Interpreting with a graph

Tests of significance

Back in the olden days, salesmen would get up on a stage and make outrageous claims about their new elixir *Oil d' Snake*. "It grows hair on the bald, takes weight off of fat people and puts weight on skinny people, and cures whatever it is that ails you… my friend," these salesmen would claim. But, where's the proof? Where…Mr. Snaky-man? Loser!

That's what a ***test of significance does***: it tests the data from a sample to see if the statistics were ***significantly*** different from the ***stated*** population mean, or the ***claim***. But remember that there will always be a natural bit of variance between a sample and a population.

How much difference does there have to be before we change our minds about a claim? At what point do we start trusting the data that has been calculated from a sample? This idea of "**how much of a difference**" can be quantified and research can be carried out through mathematical calculations. This mathematical process is the very foundation of the science of statistics! Oh yeah, feel the math! Can you? Mgz can.

In this chapter we will walk you through the basics of hypothesis testing. In a nutshell, it works like this: the ***claim***, or the ***stated*** information, (usually we refer to it as the population mean) is tested using the descriptive statistics calculated from a sample of the population. If the statistics are **too** different from the stated population mean, then we say that the data is ***significant***, or that is has ***significance***. And if we have significance from our data, we can reject the original claim for *probably* being untrue.

> **Vocabulary**
>
> **Claim**
> A claim is a statement that is either true or false.
>
> The claim is what is being tested with the null hypothesis.

If the data has been proven to be significant, then you have to act! Remember this: to have significance is saying, "There is a serious difference between what you are claiming and from what I have calculated from a sample," or, "If the original claim were true, then the results that we got from our sample would only be true one chance in twenty, but we got them on the first try. Explain that, eh?"

*Note: Read the problems carefully to determine which claim is being tested.

Null hypothesis

All statements, or claims, are either true or false. When testing a claim, a statement, or a population parameter, with statistics calculated from a sample, we label the claim as the ***null hypothesis***. We treat the null hypothesis as if it were innocent until proven guilty. We always presume that the claim is true until we find evidence of a statistical nature that proves a ***significant*** difference. If we do find a significant difference then we reject the null hypothesis. And rejection hurts!

The ***test of significance*** measures the strength, using quantitative measurements (numbers), of the case ***against*** the null being true. Normally, ***the null hypothesis states that there is no difference, or change, between the statement being tested and the sample***.

A different claim (other than the null) that we are trying to prove true for the entire population is called the ***alternative hypothesis***. If we can prove that the null hypothesis as false, or rather find that there is not enough evidence to prove it true, then the alternative hypothesis can be accepted as true. Do you see how this works? Pretty cool, don't you think?

Vocabulary

Null hypothesis (H_0)

All null hypotheses are basically stating the same thing: there is no difference, or change, between the sample and the population.

Alternative hypothesis (H_a)

All alternative hypotheses are basically stating the same thing: there is a significant difference, or change, between the sample and the population.

We write the null hypothesis as $\boldsymbol{H_0}$ and the alternative hypothesis as $\boldsymbol{H_a}$. We set up these tests of significance in a special order that will help us mathematically check the significance.

Sometimes it is easier to start with the H_a. What are we seeking evidence ***for***? Then make use of that information to figure out what the H_0 is stating.

How we begin these tests of significance is by first writing out the null and the alternative hypotheses. We state that the null hypothesis (the claim) as a number or a value, and then state with the alternative hypothesis that the true population mean is either ***less than***, ***greater than***, or ***not equal to*** the stated value from the null. These are the only three options in a test of significance.

Here is an example of a null hypothesis derived from a claim.

Example 9.1: Charlie's Chicken and Eggs Company has been selling eggs for over 25 years and knows that the mean weight of one of their eggs is 32.7 grams with a standard deviation of 4.3 grams. In order to save money they switched the hens' feed to a cheaper brand. After two months Charlie grabs 50 eggs at random and weighs them. The sample mean was 31.2 grams,

leading Charlie to believe that the new feed is making his hens lay smaller eggs. **What is the null hypothesis? What is the alternative hypothesis?**

Since Charlie has been keeping records of his eggs' weights for years and years, we will assume his mean weight is the true population mean.

This is how we write out the null hypothesis: $\boldsymbol{H_0 : \mu = 32.7}$.

From this egg-xample, there are two different options for the alternative hypothesis:

Option 1: We can say that the average weight of the eggs is **less than** 32.7 grams. This alternative hypothesis looks like this: $\boldsymbol{H_a : \mu < 32.7}$.

Option 2: We can say that the average weight of the eggs is **not equal to** 32.7 grams. This alternative hypothesis looks like this: $\boldsymbol{H_a : \mu \neq 32.7}$.

We could state that $\boldsymbol{H_a : \mu < 32.7}$, or $\boldsymbol{H_a : \mu \neq 32.7}$ because Charlie's average weight from his sample of 50 eggs was 31.2 which is both less than 32.7 and not equal to 32.7.

Null hypothesis
(H_0)

Alternative hypothesis
(H_a)

H_0: u = some number
H_a: u < same number

or

H_0: u = some number
H_a: u > same number

or

H_0: u = some number
H_a: u ≠ same number

*Important notes about hypothesis testing:

- We never try to prove the alternative hypothesis as true, only that the null is not true.
- Please oh please, notice that when setting up the null and the alternative hypotheses that they are both nearly identical except the **equality** sign (<, > or ≠) of the $\boldsymbol{H_a}$.
- ***Common student error*** - never ever use the sample mean in the alternative hypothesis! That is bad statistics and just plain wrong.

If we are using ***standardized data***, which we will, this is how we write out the null and alternative hypotheses:

Null $\boldsymbol{H_0 : \mu = 0}$

Alternative $\boldsymbol{H_a : \mu > 0}$ **or** $\boldsymbol{H_a : \mu < 0}$ **or** $\boldsymbol{H_a : \mu \neq 0}$

Note: Of course you remember that the mean of a ***standardized normal distribution* is... ***zero***, don't you? Power to the zero!

Test statistics

In order to measure the strength of a case against the null hypothesis, we need to compare the statistics gathered from a sample to the population parameters. Most of this process has already been covered in chapters 3 and 7 in this course.

We start by using the ***standardized score, or z score***, formula to calculate the ***z score*** from the sample. The number that is calculated from the standardized formulas is called the ***test statistic***. (There are other test statistics besides the z score, i.e., t test statistic, F test statistic, both of which use different formulas.)

Here is the z test statistic formula:

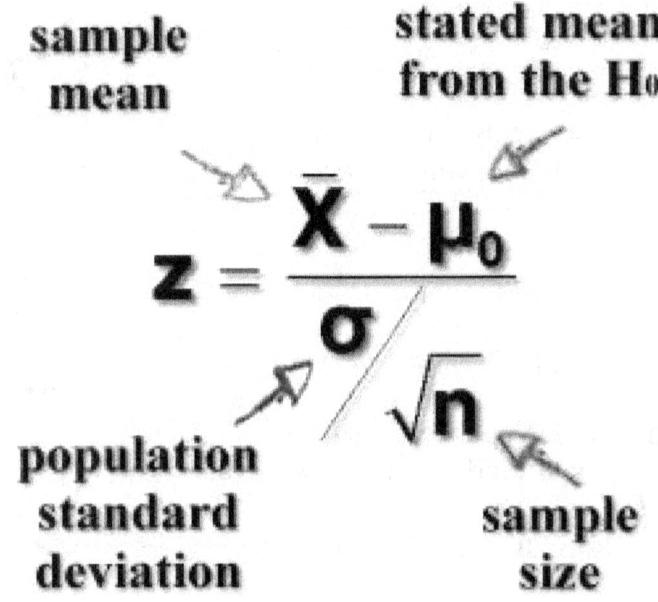

$$z = \frac{\bar{X} - \mu_0}{\sigma / \sqrt{n}}$$

*Note: The test statistic derived from this formula is based on the assumption that the null hypothesis is true.

This is best egg-xemplified with a worked example, so let's do one.

Example 9.2: Charlie's Chicken and Eggs Company has been selling eggs for over 25 years and knows that the mean weight of one of their eggs is 32.7 grams with a standard deviation of 4.3 grams. In order to save money, they switched the hen's feed to a cheaper brand. After two months Charlie grabs 50

eggs at random and weighs them. The sample mean was 31.2 grams, leading Charlie to believe that the new feed is causing his hens lay smaller eggs. **What is the test statistic from this data?**

Step 1) Use the test statistic formula:

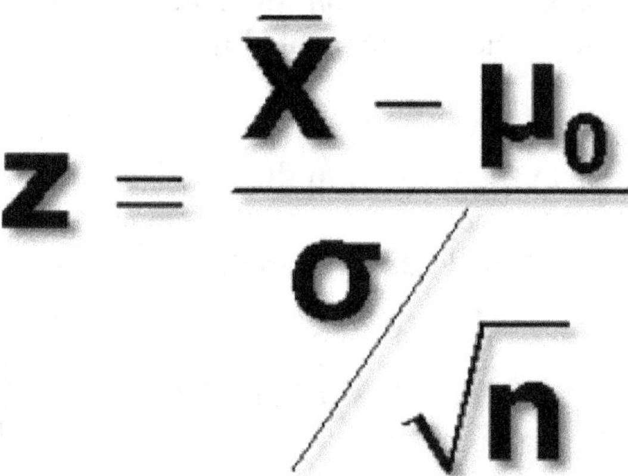

Vocabulary

Test statistic

A test statistic is a number that is calculated from a formula.

We use test statistics to look up probabilities.

Step 2) Write out the known values of the variables and then substitute them into the formula:

Sample mean **(x-bar) = 31.2**

Population mean **(μ_0) = 32.7**

Population standard deviation **(σ) = 4.3**

Sample size **(n) = 50**

Substituting, **z = (31.2 - 32.7)/(4.3/√50)**

Step 3) Math time: **z = (-1.5)/(4.3/7.1) = (-1.5)/(.61) = -2.47**

Step 4) So, the test statistic is -2.47. In other words, this result is 2.47 standard deviations **less than the mean** if the null hypothesis were true. (Remember the Empirical Rule? ***Any z score greater than 2 (or less than -2) is normally going to be significant!***

P-values

After we calculate the test statistic from the formula,

$$z = \frac{\bar{X} - \mu_0}{\sigma / \sqrt{n}}$$

we then look up its corresponding probability value, or ***p-value***, under the standardized normal distribution curve. This is the same process that we did in chapters 3 and 7: first calculate a standardized score from the data given and then find the corresponding area under the normal probability curve.

Remember the same rules apply to finding the probability from the z table:

1) If it is a ***less than*** problem (Ha < SOME NUMBER) the area in the table is the probability.

2) If it is a ***greater than*** problem (Ha > SOME NUMBER) ***subtract the area in the table from 1*** and that is the probability.

3) If it is a ***not equal to*** problem (Ha ≠ SOME NUMBER) ***pretend it is a greater than the absolute value of z problem and multiply that probability by 2*** and that is the final probability.

p-value

A small probability value is strong evidence ***against*** the null hypothesis.

A ***small p-value is strong evidence against the null hypothesis being true***. A small p-value translates into this: if the null hypothesis were true, then we would expect to get these results from a random sample would be roughly one time out of twenty trials, but we got these results on the very first try. This doesn't make a lot of sense **if the null were true to begin with**, does it? But, be careful because it could have indeed happened by chance. Remember that we, as statisticians, must expect extreme results like this 5% of the time. That's the way chance works, don't you know?

*Note: Don't forget that if the H_a has a ***greater than*** symbol, you need to ***subtract*** the p-value from the normal distribution table of the z test statistic ***from 1***.

Let's continue with Charlie's egg test and find the calculated p-value, shall we?

Example 9.3: Charlie's Chicken and Eggs Company has been selling eggs for over 25 years and knows that the mean weight of one of their eggs is 32.7 grams with a standard deviation of 4.3 grams. In order to save money they switched the hen's feed to a cheaper brand. After two months Charlie grabs 50

eggs at random and weighs them. The sample mean was 31.2 grams, leading Charlie to believe that the new feed is making his hens lay smaller eggs. **What is the p-value from this data?**

Step 1) Find the test statistic with the formula:

$$z = (\bar{X} - \mu)/(\sigma/\sqrt{n}) = (31.2 - 32.7)/(4.3/\sqrt{50}) = -2.47$$

Step 2) Look up the corresponding p-value in the normal standardized distribution table, or z table: z -2.47 >>> **.0068.**

Step 3) Interpret what this means: *if the null were true and Charlie repeated this random sampling of 50 eggs 10,000 times, then he should get this same sample mean around 68 times.* But he got it on the first try! So, the null hypothesis is probably no longer true. In other words, the hens are laying smaller eggs.

All Ha statements say the same thing: there is a difference between any of the different groups.

P-values for alternative hypotheses that are **NOT EQUAL TO**.

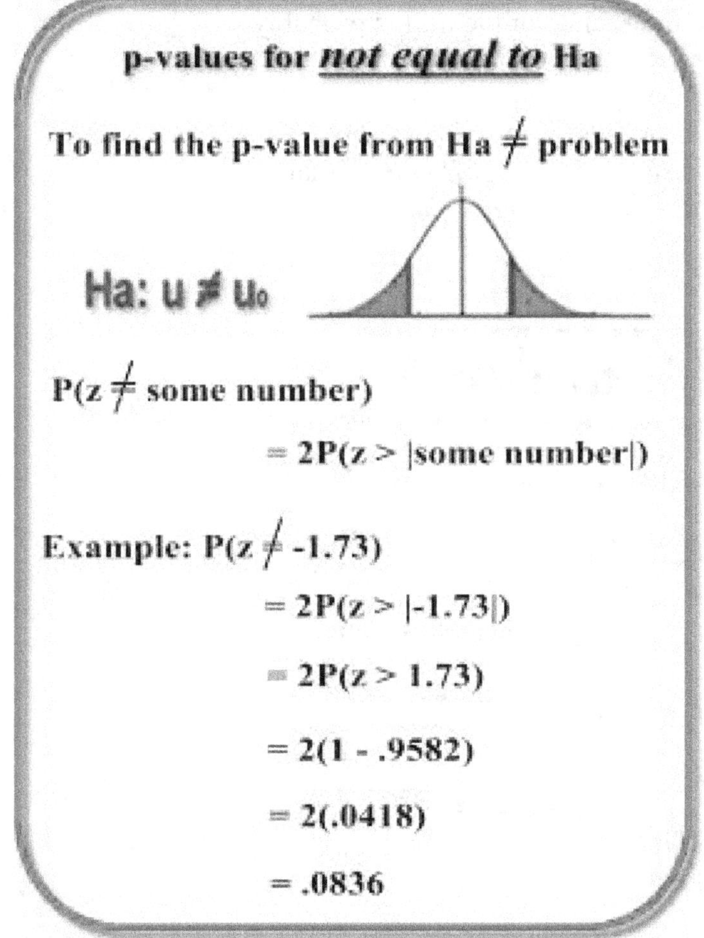

All Ho statements say the same thing: there is no difference between any of the different groups.

Significance

A small probability value is strong evidence ***against*** the null hypothesis.

Small probability values are strong evidence against the null. But, how small do the p-values have to be before we can officially declare that the null is not acceptable? These cut-off p-values are normally left up to the researcher, but if not directly stated, then the cut-off (critical) p-value is .05. Scientists from around the world have agreed that weird extremes naturally happen around 1 time in 20, and are to be expected.

In hypothesis testing we compare the **calculated probability value** to a **preset cut-off probability value** to determine if the results from the sample fall within an acceptable range: the ***confidence interval***! This cut-off value is called ***alpha (α)*** and is normally .05 unless otherwise stated. Alpha is also called a ***critical value*** because it is … a … "critical" … value: if the ***calculated*** p-value is **<u>less than</u>** the ***critical*** value, you must get "critical" of the null and ***reject*** it! If not, then we ***fail to reject***.

Another way to make your decision whether to reject the null or not is to use the confidence interval. ***If the calculated p-value falls within the confidence interval then the null is probably true***, but if the calculated p-value falls outside of the confidence interval, then the null is probably not true, and therefore we can ***flush*** (reject) the null hypothesis!

If we have enough evidence from our calculations to reject the null hypothesis, then we say that the data is statistically ***significant*** at that specific alpha level. We can also say that we have ***significance***.

reject the null or not?

If critical p-value > calculated p-value

<u>Reject the null!</u>

If critical p-value < calculated p-value

<u>Do not reject the null!</u>

Example:
If alpha = .05 and the calculated p-value = .03
then .05 > .03 tells us to
reject the null

Here are a few examples to drive this idea home.

Example 9.4: If the critical p-value (α) was .01 and the calculated p-value was 0.14, then the data is not statistically significant at the alpha level of .01. We do not reject the null.

Example 9.5: If the critical value (α) was .05 and the calculated p-value was 0.04, then the data is statistically significant at the alpha level of .05, and we reject the null.

Now it's time to put all of these steps together and perform a significance test on Charlie's egg research.

Example 9.6: Charlie's Chicken and Eggs Company has been selling eggs for over 25 years and knows that the mean weight of one of their eggs is 32.7 grams with a standard deviation of 4.3 grams. In order to save money they switched the hen's feed to a cheaper brand. After two months Charlie grabs 50 eggs at random and weighs them. The sample mean was 31.2 grams, leading Charlie to believe that the new feed is making his hens lay ***smaller*** eggs. **Is he correct? (Reject or do not reject the null?)**

Step 1) Write out the null and alternative hypothesis:

$\mathbf{H_0: \mu = 32.7 \quad H_a: \mu < 32.7}$

Step 2) Find the test statistic with the formula:

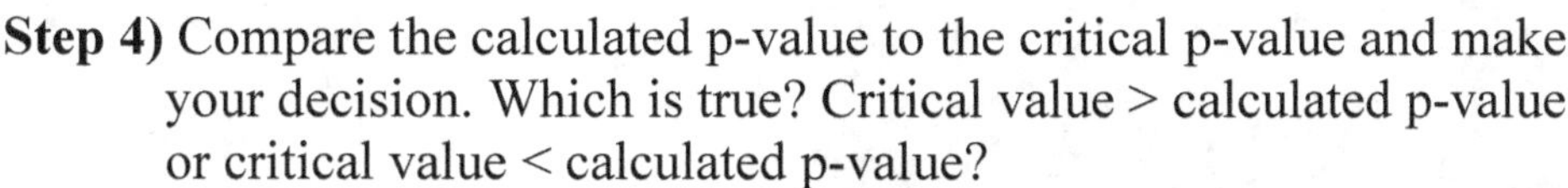

$$z = (\bar{X} - \mu)/(\sigma/\sqrt{n}) = (31.2 - 32.7)/(4.3/\sqrt{50}) = -2.47$$

Step 3) Look up the corresponding p-value in the normal standardized distribution table, or z table: z -2.47 >>> **.0068.**

Step 4) Compare the calculated p-value to the critical p-value and make your decision. Which is true? Critical value > calculated p-value, or critical value < calculated p-value?

.05 (critical value or α) > .0069 (calculated p-value from test statistic) is true and therefore we **reject the null** and say that this data is **significant at the .05 level**. So, according to Charlie's research, his hens are indeed laying smaller eggs.

Type I and Type II errors

There are two types of errors in testing for significance. ***Type I error***, also known as a false positive, is the ***error of rejecting a null hypothesis when it is actually true***. The null always states that there was no significant difference between the groups, so an example of a Type I error would be to say that there is a (significant) difference between the groups when actually there is no (significant) difference.

Type I error

Type I error is rejecting the null hypothesis when it is true.

Another Type I error example would be if a pregnancy test shows that a woman is pregnant when in reality she is not. (The null would be claiming that she is not pregnant.)

Type II error, also known as a false negative, is the ***error of failing to reject a null hypothesis when it is not true***. The null always states that there was no difference between the groups, so an example of a Type II error would be to say that there is no (significant) difference between the groups when actually there is (significant) difference.

Another Type II error example would be if a pregnancy test shows that a woman is not pregnant when in reality she is.

Here is a table that explains the types of errors:

Statistics - Hypothesis Test

	Null Hypothesis True	Null Hypothesis False
Reject Null Hypothesis	Type I Error	Correct
Fail to Reject Null Hypothesis	Correct	Type II Error

Test of significance - the entire process

Hypothesis testing

1) Write out Ho and Ha
2) Find the test statistic
3) Find the p-value
4) Compare the calculated p-value to the critical p-value
5) Make your decision: reject the null or not

Let's do a few examples of these tests for significance using the following steps.

Step 1) Write out the null and alternate hypotheses.

Step 2) Find the test statistic.

Step 3) Find the corresponding probability value (the area under the curve) to the test statistic either from the standardized normal distribution curve, or from Excel.

Step 4) Compare the **calculated p-value** to the **critical p-value**. (Which is larger?)

Step 5) Either reject the null hypothesis, or not, based on the data.

Step 6) Reward yourself! Go ahead, you deserve it!

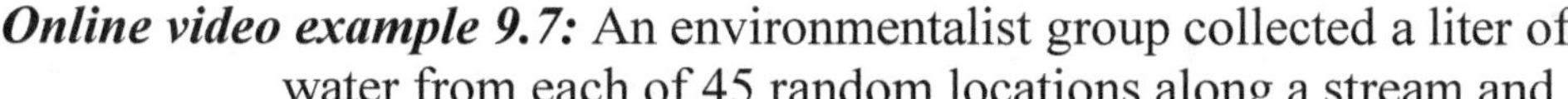

Online video example 9.7: An environmentalist group collected a liter of water from each of 45 random locations along a stream and measured the amount of dissolved oxygen in each specimen. The mean from their samples is 4.62 milligrams (mg). Is this strong evidence that the stream has a mean oxygen content of less than 5 mg per liter? (Suppose we know that dissolved oxygen varies among locations according to a normal distribution with $\sigma = 0.92$ mg.)

Step 1) $\mathbf{H_0: \mu = 5 \quad H_a: \mu < 5}$

Step 2) Sample mean = 4.62, population mean = 5, population standard deviation = 0.92 and sample size = 45

$$\mathbf{z = (4.62 - 5)/(.92/\sqrt{45}) = -2.77}$$

Step 3) z of -2.77 generates a p-value of **.0028**

Step 4) Calculated p-value (.0028) < critical p-value (.05)

Step 5) We have a significant difference therefore we **reject the null** hypothesis. The stream has an oxygen content of less than 5 mg per liter.

Step 6) Break time! Mgz will be back in a minute... Ahhh! I am back. Miss me?

Online video example 9.8: In 2005, California state 11th grade test scores averaged showed that 95.6% of all California 11th graders scored "less than proficient" in algebra. The average test score was 27 correct out of 72 multiple choice questions with a standard deviation of 11.6. The new Superintendent of Education makes all kinds of changes in the teaching practices of algebra. Two years later he has 100 11th graders chosen at random take the same test and they had an average of 30 correct. The superintendent crows like a rooster and states that his program changes do work better and the proof is in the new test scores. Prove him right or wrong.

Step 1) H_0: $\mu = 27$ H_a: $\mu > 27$

Step 2) Sample mean = 30, population mean = 27, population standard deviation = 11.6 and sample size = 100

$$z = (30 - 27)/(11.6/\sqrt{100}) = 2.59$$

Step 3) z of 2.59 generates a p-value of **.9952, but we need to subtract this from 1 because it is a *<u>greater than</u>* problem:**
1 - .9952 = .0048

Step 4) Calculated p-value (.0048) < critical p-value (.05)

Step 5) We have a significant difference therefore **we reject the null** hypothesis. Therefore the students' mean scores are higher than 27. We will assume that it was because of the superintendent's changes.

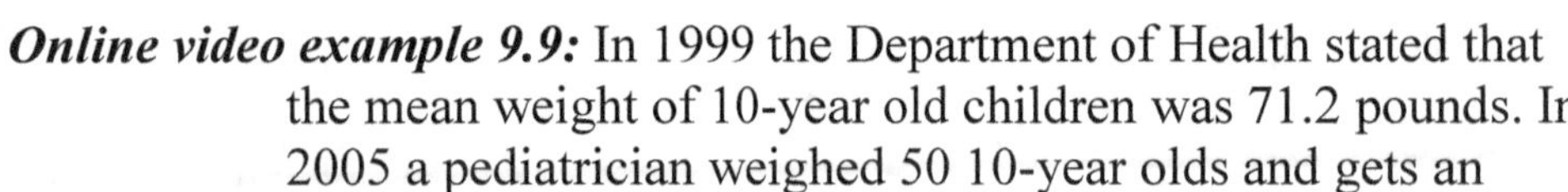

Online video example 9.9: In 1999 the Department of Health stated that the mean weight of 10-year old children was 71.2 pounds. In 2005 a pediatrician weighed 50 10-year olds and gets an

average weight of 73.0 pounds with a standard deviation of 8.3 pounds. The doctor claims that the mean weight of 10-year olds is not 71.2. Prove or disprove his claim. ($\alpha = 0.05$)

Step 1) H_0: $\mu = 71.2$ H_a: $\mu \neq 71.2$

Step 2) Sample mean = 73, population mean = 71.2, population standard deviation = 8.3 and sample size = 50

$$z = (73 - 71.2)/(8.3/\sqrt{50}) = 1.53$$

Step 3) z of 1.53 generates a p-value of **.9370, but we need to subtract this from 1 and then multiply that by 2 because it is a *<u>not equal to</u>* problem: 1 - .9370 = .0630 …
.0630 x 2 = .1260**

Step 4) Calculated p-value (.1260) > critical p-value (.05)

Step 5) We do not have a significant difference therefore we **do not reject the null** hypothesis. The doctor did not find enough evidence to prove that the mean weight of 10-year olds was not 71.2 pounds.

Test of significance – test statistics versus p-values

***This is an important note: the "less than" or "greater than" rejecting-the-null rules when comparing your calculated p-value against the critical p-value depends on the individual statistics teacher. If the teacher is using test statistics (i.e., z score, t score, F ratio) to compare data instead of p-values, then remember that a LARGE test statistic (ex: z score of 2.74) is strong evidence against the null.

Important!

Large test statistics generate small p-values!

But if your teacher is using probability values (p-values), then always remember that a ***SMALL p-value is strong evidence against the null.***

These examples will help to explain this idea.

Online video example 9.10: If you have a calculated p-value of .08, you would not reject the null at the .05 level.

Online video example 9.11: If you have a calculated z score of 2.15, you would reject the null at the .05 level (because the critical value that generates .05 is a z score of 1.960).

Online video example 9.12: If you have a calculated p-value of .006, you would reject the null at the .01 level.

Online video example 9.13: If you have a calculated z score of 2.15, you would not reject the null at the .01 level (because the critical value that generates .01 is a z score of 2.576).

test statistics or p-values?

using probablity values

critical value > calculated p-value

REJECT the null

---

using test statistics

(i.e., z score, t score, F ratio)

critical value < calculated test statistic

REJECT the null

Important!

Large test statistics generate small p-values!

Chapter 9 Review

Tests of significance are a way to look for evidence calculated from the sample data that supports, or does not support, a claim about a population parameter, usually the mean. The statement being tested is called the ***null hypothesis***.

Chapter Review

The null hypothesis is the ***claim***, or ***stated value***, that is being tested. All null hypotheses say the same thing: there is no significant difference between any of the groups.

A ***test statistic*** is the number that is calculated from the standardized formula. Test statistics generate specific ***probability values (p-values)***. A ***small p-value, or a large test statistic, is strong evidence against the null hypothesis***.

If we have enough evidence from our calculations to reject the null hypothesis, then we say that the data is statistically ***significant*** at that specific alpha level. We can also say that we have ***significance***.

The entire process of hypothesis testing:

1) Write out H_0 and H_a
2) Find the test statistic
3) Find the p-value
4) Compare the calculated p-value to the critical p-value
5) Make your decision: reject the null or not

Guided homework Chapter 9: Null hypothesis

1) A candy company makes butterscotch flavored candies. The mean weight of the candies follows a normal distribution with a mean weight of 1.672 grams with a standard deviation of 0.387 grams. Arnold, the new Quality Control Technician, believes the machine is not working properly and measures the weights of 50 randomly selected candies. The average weight that he gets is 1.581 grams. He runs to the president and says that he believes that the machine is not putting enough candy mix into each candy. Does the evidence prove him correct?

   a. What is the null hypothesis?

   b. What is the alternative hypothesis?

   c. What is the test statistic?

   d. What is the P-value?

   e. Do we accept or reject the null hypothesis at the 5% level of significance? (Is it statistically significant at the 5% level? ($\alpha = .05$))

   f. Do we reject the null hypothesis at the 1% level of significance? (Is it statistically significant at the 1% level? ($\alpha = .01$))

2) An environmentalist group (carefully!) weighed 50 baby polar bears that they found randomly. The mean weight was 43.9 kilograms with a standard deviation of $\sigma = 8.5$. The environmentalists claim that this is strong evidence that the entire population of polar bears has a mean weight of less than 45.5 kilograms that was reported in a similar survey conducted in 1981.

   a. What is the null hypothesis?

   b. What is the alternative hypothesis?

   c. What is the test statistic?

   d. What is the appropriate p-value?

e. Using the significance level $\alpha = 0.1$, would you accept the null?

f. Using the significance level $\alpha = 0.05$, would you accept the null hypothesis?

3) In 1999 the Department of Health stated that the mean weight of 10-year old children was 71.2 pounds. In 2005 a pediatrician weighed 50 10-year olds and gets an average weight of 73.0 pounds with a standard deviation of 8.3 pounds. The doctor claims that the mean weight of 10-year olds is not 71.2. Prove or disprove his claim. ($\alpha = 0.05$)

4) A coffee shop manager receives a report from corporate headquarters stating that the average daily sales quotas should be a minimum of $5,692 with a standard deviation of $319. He tallied his daily receipts for the month of March (31 days) and came up with a daily average of $6,104. He calls the president of the company and insists that the company-wide minimum daily sales be increased. If you were the president, would you listen to him or fire him because he can't do the math? ¿Por qué?

5) You have a friend working at the ACLU and she calls you to ask for your crazy-mad statistics skills on a case that she is working on. She is putting together a class action suit against a major soft drink company. In the suit, the company is accused of putting the incorrect amount of carbohydrates on the label (40 grams). Your friend says she hired a private laboratory to measure the true amount of carbohydrates in one hundred 12 oz. cans of soda and they got an average of 40.12 grams. Assuming that the standard deviation is 1.08 grams, can your friend state under oath that the mean average of carbohydrates is, in fact, greater than 40 grams?

Chapter 9 Quiz

1. Assume that we are testing a hypothesis such that

H_0: u = 0 and Ha: $u \neq o$

If the test statistic was -0.84, what would the corresponding probability be?

a. .2242
b. .4009
c. .7995
d. .2005
e. 1.5991

2. What is the p-value of the test statistic of 2.06?

3. True or false?

A Type II error would be not rejecting the null hypothesis when it was actually false.

4. True or false?

When a calculated p-value is less than the critical value alpha we say that if the null hypothesis were true to begin with then we would rarely get the results from the test of significance, yet we got them on our first sample.

5. True or false?

Reject the null whenever the calculated p-value is less than the critical alpha value.

6. Peter's Pizzeria has been buying sausage from Sammy's Sausage Factory for over 15 years. Peter knows that the average weight of the type of sausage that he uses is 102 grams with a standard deviation of 15.4 grams. Peter hears that Sammy Sr. retired and that Sammy Jr. has taken over the business, so he takes out 80 sausages and weighs them and gets an average weight of 100 grams. Peter thinks that Sammy Jr. might be selling him a smaller sausage.

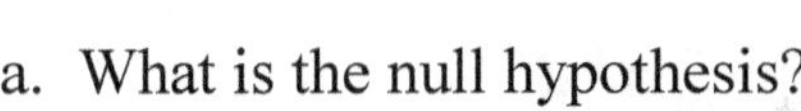

a. What is the null hypothesis?

b. What is the alternative hypothesis?

c. What is the test statistic?

d. What is the p-value of the test statistic?

e. Is the data significant at $\alpha = .05$?

f. Is the data significant at $\alpha = .01$?

7. Assume that you are testing

H0: $\mu = 4.3$ and that Ha: $\mu > 4.3$

If the test statistic was 1.97, what is the corresponding p-value?

8. True or false?

A p-value of .04 compared to a critical value alpha of .05 would tell you to reject the null.

9. True or false?

You would not reject the null at the .01 level from a test of

H0: u = 0 and Ha: u $\neq$ 0

with a test statistic of 2.66.

10. Full grown female elephants have an average weight of 5400 kgs with a standard deviation of 522 kgs. A naturalist weighs 100 full-grown female elephants (which wasn't easy) and gets a mean weight of 5310 kgs. The naturalist states that elephants are losing weight.

a. What is the alternative hypothesis?

b. What is the null hypothesis?

c. What is the test statistic?

d. What is the p-value of the test statistic?

e. Is the data significant at $\alpha = .05$?

f. Is the data significant at $\alpha = .01$?

11. What would the test statistic be from a sample size of 60 with a sample mean of 269 if a population's distribution was stated as N(267, 43.9)?

12. True or false?

The null and the alternative hypotheses are written out nearly identically except for the equality symbol between the μ and the number.

13. True or false?

A test of significance has a null hypothesis and a responsive hypothesis.

14. True or false?

A student tests negative for the Swine Flu when in reality she does have the flu is an example of a Type I error.

15. True or false? - The test statistic is the same as the probability value.

16. Assume that we are testing a hypothesis such that $\mathbf{H_0}$: $\mu = 0$ and $\mathbf{H_a}$: $\mu \neq 0$. If the test statistic was 1.55, what would the corresponding probability be?

17. If there is enough evidence to reject the null hypothesis we call the data ______ at the alpha level.

a. proof
b. binomial
c. explanatory
d. significant
e. probable

18. What two purposes does a confidence interval have?

a. Estimate the true population mean and to test for significance
b. To develop a binomial setting and to find the probability
c. To find the sample size and to the check the distribution for symmetry
d. To calculate general probability and to find the sample mean
e. To find the population standard deviation and to find the sample mean

19. True or false? - A Type I error would be rejecting the null hypothesis when it was actually true.

20. True or false? - A student tests positive for the Swine Flu when in reality she does not have the flu is an example of a Type I error.

21. True or false?

Testing a statement for accuracy with statistics is called testing the null hypothesis.

Ch 10: t tests

Chapter

10

t tests

Hypothesis testing is much more accurate with large sample sizes, as the Central Limit Theory and the Law of Large Numbers tell us.

These statistical formulas can only be used if we already know the true population standard deviation. But if you think about that for a while you should be asking yourself, "If we don't know the true population mean, how are we supposed to know what the true population standard deviation is? After all, don't we use the mean to find the standard deviation in the first place?" The answer is yes, we do. Good thinking!

When we don't know the true population standard deviation, we use a ***t test*** to check for significance. These ***t tests*** are very similar to the process of ***z score testing*** but use slightly different formulas and a different distribution table that has built-in flexibility using what are called ***degrees of freedom***.

In this chapter, we will investigate ***t tests***:

- The different formulas for a ***t test***
- One-sample ***t test*** statistic
- Confidence interval for a ***t*** distribution
- Critical values for ***t***
- One-sample ***t test***: the entire process
- Two-sample ***t test***

What's in this chapter

- The different formulas for a t test
- One-sample t test statistic
- Confidence interval for a t distribution
- Critical values for t
- One-sample t test: the entire process
- Two-sample t tests

t tests formulas and when to use them

There are two types of ***t tests***. A ***one-sample t test*** is used when comparing the means between a small sample size (less than or equal to 30) against a stated parameter, usually the population mean, or if the population standard deviation is unknown. A ***two-sample t test*** is used to compare the differences between two different samples (regardless of sample size) from the same population. All the rules of a normal distribution must apply or your ***t test*** results will be invalid.

t test

Use a t test:

- If the population standard deviation is unknown
- If the sample size is thirty or less
- If two samples from the same population (regardless of sample size) are being compared

Any study that has at least one of the following three conditions should use a t test:

Condition #1) If the population standard deviation is unknown (***one-sample t test***).

Condition #2) If the sample size is thirty or less (***one-sample t test***).

Condition #3) If two samples from the same population (regardless of sample size) are being compared (***two-sample t test***).

*Note: ***t tests*** are only used to compare the means of two different groups of data. If more than two groups are to be compared, we must use an ***analysis of variance (ANOVA).*** ANOVAs are usually the first thing taught in a graduate statistics class.

When using a t test because ***the population's standard deviation is not known***, we must use the sample's standard deviation. When we use the sample's standard deviation we call it the ***standard error***. The slight difference between a standard deviation and a standard error is that the sample's ***standard error is the standard deviation divided by the square root of the sample size***!

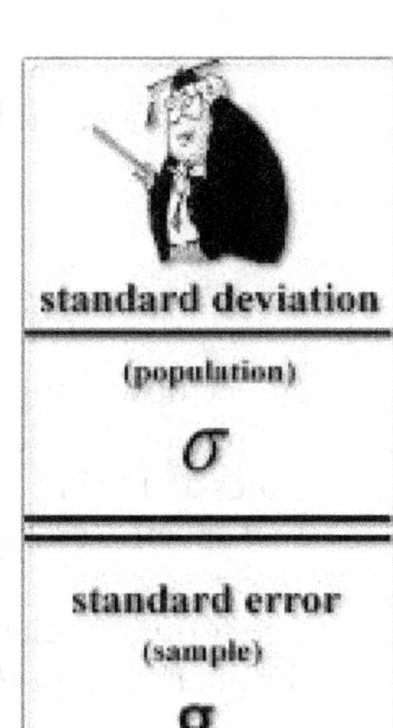

or

$\frac{S}{\sqrt{n}}$

Example 10.1: A sample size of 16 with a standard deviation of 2.8 would generate a standard error of $2.8/\sqrt{16} = 0.7$

Another difference between a z test and a t test are the different critical values generated by the two different distributions. The t distribution table is different from the z distribution table and takes a little practice to find the associated p-values.

Remember the *z* ***test statistic*** formulas?

Well, here are the ***one-sample t test statistic*** formulas:

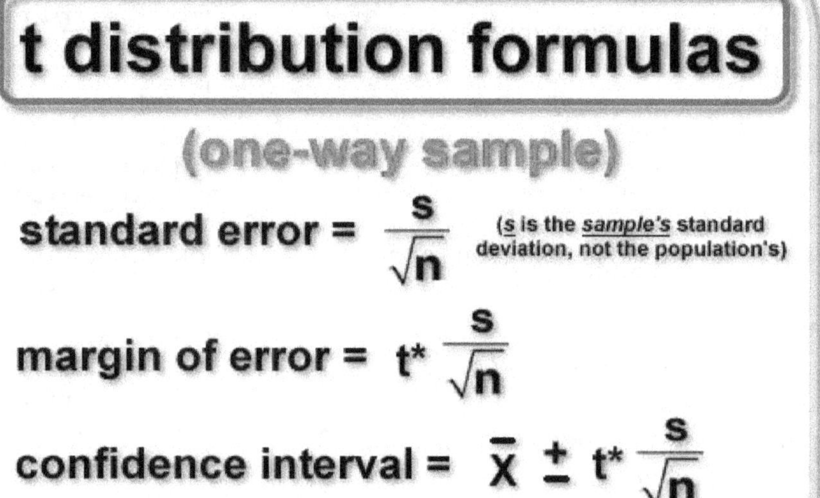

In the confidence interval formula for a t test, t* represents the critical value (from a table) corresponding to a a level of confidence.

Notice how similar the formulas are? The z and the t distribution formulas are practically identical, but their normal distribution tables are not. The z tables and the t tables are set up differently and have a different process to find specific probabilities.

The formulas related to the one-sample t test are the same as the z test formulas, except ***that z* is substituted with t*.***

Note: The confidence interval generated by the t critical value is the area under the normal distribution curve between -t* and t*. There are no negative t test statistics because we use the ***absolute value of t.***

Remember from chapter 8 how the *z** (the critical value from the z table) depends on the level of confidence C? Example: a 90% confidence interval corresponds to a z* value of 1.645, which is used in the margin of error and the confidence interval formulas.

Well, the ***t**** (a different critical value from the ***t distribution table***) is found on the t table by the number of its ***degrees of freedom*** and by its ***level of confidence C***.

Degrees of freedom

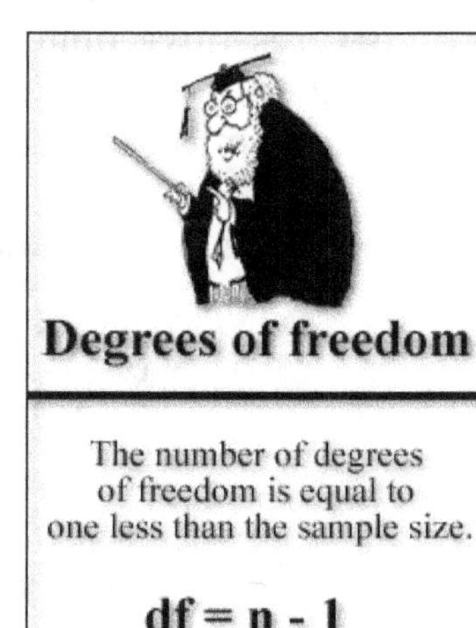

Degrees of freedom are a difficult concept to explain, but how is this? The degrees of freedom of a calculation is equal to the number of independent measurements (***n***) that go into the calculation minus the number of parameters that we are estimating (normally just 1: the mean).

What you have to remember is that the number of degrees of freedom is ***normally*** equal to ***one less that the sample size***.

Degrees of freedom ***(df) = n - 1***

The t distribution curve is similar to the z distribution curve in shape, but the ***fewer*** the degrees of freedom, the "shorter" and "wider" the curve becomes; t distributions have more area in their tails than a z curve.

The more degrees of freedom, the more the t distribution approaches the standard normal distribution, the z distribution. In fact, ***anything greater than 29 degrees of freedom (n > 30) is to be treated as a z distribution***.

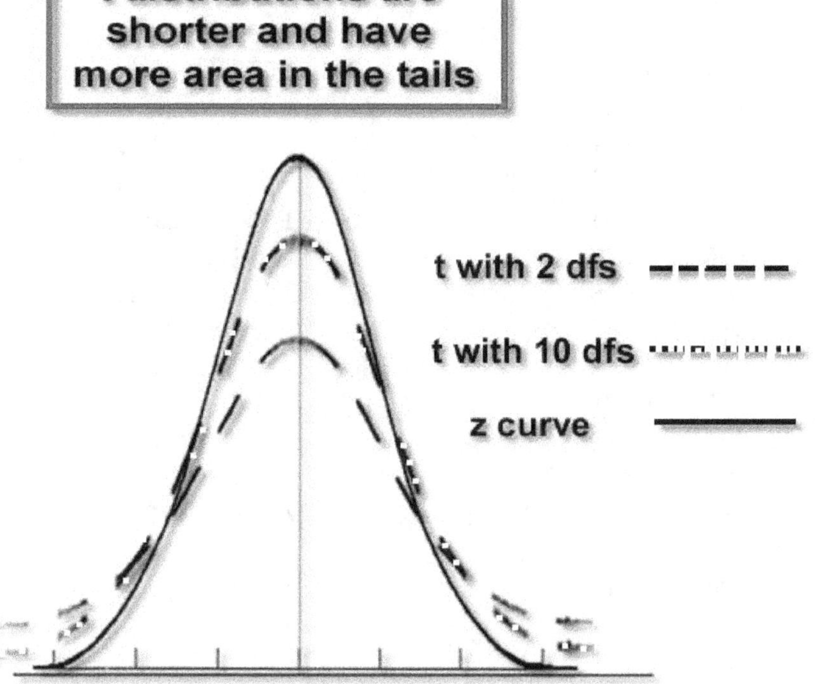

t distribution table

The ***critical value t**** is found on the t table where the appropriate **degrees of freedom row** (on the left) ***intersects*** with the corresponding **level of confidence C** (column at the top). Remember that the number of degrees of freedom are one less that the sample size.

Here is the ***t distribution table***:

t distribution critical values table

	Level of confidence, c	0.50	0.80	0.90	0.95	0.98	0.99
	One tail, α	0.25	0.10	0.05	0.025	0.01	0.005
d.f.	Two tails, α	0.50	0.20	0.10	0.05	0.02	0.01
1		1.000	3.078	6.314	12.706	31.821	63.657
2		.816	1.886	2.920	4.303	6.965	9.925
3		.765	1.638	2.353	3.182	4.541	5.841
4		.741	1.533	2.132	2.776	3.747	4.604
5		.727	1.476	2.015	2.571	3.365	4.032
6		.718	1.440	1.943	2.447	3.143	3.707
7		.711	1.415	1.895	2.365	2.998	3.499
8		.706	1.397	1.860	2.306	2.896	3.355
9		.703	1.383	1.833	2.262	2.821	3.250
10		.700	1.372	1.812	2.228	2.764	3.169
11		.697	1.363	1.796	2.201	2.718	3.106
12		.695	1.356	1.782	2.179	2.681	3.055
13		.694	1.350	1.771	2.160	2.650	3.012
14		.692	1.345	1.761	2.145	2.624	2.977
15		.691	1.341	1.753	2.131	2.602	2.947
16		.690	1.337	1.746	2.120	2.583	2.921
17		.689	1.333	1.740	2.110	2.567	2.898
18		.688	1.330	1.734	2.101	2.552	2.878
19		.688	1.328	1.729	2.093	2.539	2.861
20		.687	1.325	1.725	2.086	2.528	2.845
21		.686	1.323	1.721	2.080	2.518	2.831
22		.686	1.321	1.717	2.074	2.508	2.819
23		.685	1.319	1.714	2.069	2.500	2.807
24		.685	1.318	1.711	2.064	2.492	2.797
25		.684	1.316	1.708	2.060	2.485	2.787
26		.684	1.315	1.706	2.056	2.479	2.779
27		.684	1.314	1.703	2.052	2.473	2.771
28		.683	1.313	1.701	2.048	2.467	2.763
29		.683	1.311	1.699	2.045	2.462	2.756
∞		.674	1.282	1.645	1.960	2.326	2.576

One-sample t test statistic

The t test statistic is similar to the z test statistic in that it tells us how far away any measurement is from the mean in a standardized distribution. But ***sample size plays a huge role in the t test statistic***.

*Note: A one-sample t test is also known as a ***student's t test***.

Here is the formula for a one-way t test statistic: (**Note: the μ refers to the stated population mean from the null hypothesis. This one-sample t test compares a small sample against a stated claim, normally the population mean.)

$$t = \frac{\bar{X} - \mu}{\sigma / \sqrt{n}}$$

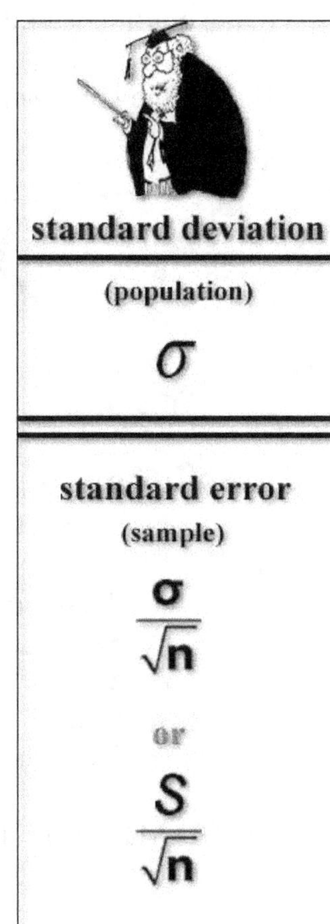

The significance testing of a t score follows the same procedures as a z test. Let's do an example.

Example 10.2: Sra. Sanchez gave her 11th grade algebra class a test from the SAT. She has 14 students in her class and they scored an average of 518 with a standard deviation of 28. She knows that the national average score for the SAT algebra test is 500. **What is the *t test statistic* from her class scores compared to the national average?**

Step 1) Make sure this a t test situation: we don't know the population standard deviation and the sample size is less than 30, so yes, this is a t test problem. (***Use the sample's standard deviation in the formula***.)

Step 2) Substitute the known values into the t test statistic formula:

t = (518 - 500)/(28/√14)

Step 3) Math time, oh boy!: **t = (518 - 500)/(28/√14)** = *2.40*

Step 4) Interpret what this means*: If many many different groups of 14 11th graders took the same test under the same conditions, then Sra. Sanchez' class would be 2.40 standard deviations above the mean (norm). That's a huge difference! Nice job Sra. Sanchez!*

One-sample t test confidence interval

The confidence level for a t distribution is similar to a z test, but the critical values come from the t table instead of the z table. Remember that a confidence interval is the sample mean plus/minus a margin of error.

Here is the ***confidence interval*** formula for a t test:

$$\bar{X} \pm t^* \frac{s}{\sqrt{n}}$$

In the confidence interval formula for a t test, t* represents the critical value (from a table) corresponding to a a level of confidence.

Example 10.3: Sra. Sanchez gave her 11th grade algebra class a test from the SAT. She has 14 students in her class and they scored an average of 518 with a standard deviation of 28. She knows that the national average score for the SAT algebra test is 500. Sra. Sanchez wants to check the validity of the stated national average score of 500 and wants to build a 90% confidence interval for the true population mean. **What is a 90% CI from this data?**

Step 1) Make sure this a t test situation: we don't know the population standard deviation and the sample size is less than 30, so yes, this is a t test problem.

Step 2) Substitute the known values into the t test statistic formula:

$$\mathbf{CI = 518 \pm t^*(\mathit{s}/\sqrt{\mathit{n}})}$$

Step 3) Figure out what the standard error is: **s/√n = 28/√14 =** ***3.74***

Step 4) ***<u>Here comes the new part</u>***... find the t* value from the t distribution table:

a) Find the degrees of freedom = **n - 1 = 14 - 1 = 13**.

b) In the t table, find where the ***13th df row intersects with the 90% level of confidence C column***. From the table below it looks like **t* = 1.771.**

t distribution critical values table

d.f.							
	Level of confidence, c	0.50	0.80	0.90	0.95	0.98	0.99
	One tail, α	0.25	0.10	0.05	0.025	0.01	0.005
d.f.	Two tails, α	0.50	0.20	0.10	0.05	0.02	0.01
1		1.000	3.078	6.314	12.706	31.821	63.657
2		.816	1.886	2.920	4.303	6.965	9.925
3		.765	1.638	2.353	3.182	4.541	5.841
4		.741	1.533	2.132	2.776	3.747	4.604
5		.727	1.476	2.015	2.571	3.365	4.032
6		.718	1.440	1.943	2.447	3.143	3.707
7		.711	1.415	1.895	2.365	2.998	3.499
8		.706	1.397	1.860	2.306	2.896	3.355
9		.703	1.383	1.833	2.262	2.821	3.250
10		.700	1.372	1.812	2.228	2.764	3.169
11		.697	1.363	1.796	2.201	2.718	3.106
12		.695	1.356	[illegible]	2.179	2.681	3.055
13		.694	1.350	1.771	2.160	2.650	3.012
14		.692	1.345	[illegible]	2.145	2.624	2.977
15		.691	1.341	1.753	2.131	2.602	2.947
16		.690	1.337	1.746	2.120	2.583	2.921
17		.689	1.333	1.740	2.110	2.567	2.898
18		.688	1.330	1.734	2.101	2.552	2.878
19		.688	1.328	1.729	2.093	2.539	2.861
20		.687	1.325	1.725	2.086	2.528	2.845
21		.686	1.323	1.721	2.080	2.518	2.831
22		.686	1.321	1.717	2.074	2.508	2.819
23		.685	1.319	1.714	2.069	2.500	2.807
24		.685	1.318	1.711	2.064	2.492	2.797
25		.684	1.316	1.708	2.060	2.485	2.787
26		.684	1.315	1.706	2.056	2.479	2.779
27		.684	1.314	1.703	2.052	2.473	2.771
28		.683	1.313	1.701	2.048	2.467	2.763
29		.683	1.311	1.699	2.045	2.462	2.756
∞		.674	1.282	1.645	1.960	2.326	2.576

Step 5) Substitute the know values and do some math:
CI = 518 ± ***t****(s/√n) = 518 ± ***1.771***(3.74) = **518 ± 6.62** ... or written out in interval form: **[511.38, 524.62]**

Step 6) Interpretation time: *If many many groups of fourteen 11th graders took the same test, then 90% of their mean scores should fall within this range.*

One-sample t test of significance -the entire process

Let's do a one-sample t test, okee dokee? The steps are identical to a test of significance of a null hypothesis, but with different critical values (from the t table) and standard deviations (we use the standard errors)... and that's what makes it a t test!

The steps to test for significance in a t test are as follows:

Step 1) Write out the null and alternate hypotheses.

Step 2) Find the test statistic.

Step 3) Find the corresponding probability (the area under the curve) to the t test statistic either from the t distribution curve, or from Excel.

Step 4) Compare the calculated p-value to the critical p-value. (Which is larger?)

Step 5) Either reject the null hypothesis, or not, based on the data.

Step 6) Interpret the results.

Online video example 10.4: In 1976, the National Football League stated that the average weight of professional football players that played on the line (the linemen - muy macho!) was 278 pounds. In 2004 a concerned government official was worried that too many professional athletes were using performance enhancing drugs, especially steroids. She was given permission by the NFL to do a research study on the linemen of the San Diego Chargers (go Chargers!). She weighed all of the linemen and recorded the following weights:

279	300
298	296
275	279
267	281
311	292
306	288
291	268
286	277
278	271
275	296

Is this significant evidence to claim that the linemen of 2004 were indeed larger than in 1978 at an alpha (α) level of .01?

Step 1) Write out the null and alternate hypotheses:
H_0: μ = 278, Ha: μ > 278 (The " > " symbol tells us that we are going to use a one-sided, or one-tailed, p-value.)

Step 2) Find the t test statistic: we **first need to find the sample's mean (x-bar)** and **standard deviation** (s) which we will now call the standard error (because it is from a sample, not the population):
x-bar = 285.70, s = 12.67, n = 20 ... t = (x-bar - μ)/(s/$\sqrt{n}$) = (285.70 - 278)/(12.67/$\sqrt{20}$) = 7.70/2.83 = 2.72

(*Note: We can already tell, from the Empirical Rule, that the data will be significant because of this very large test statistic. Large test statistics generate small p-values. So, act surprised when we finish this problem.)

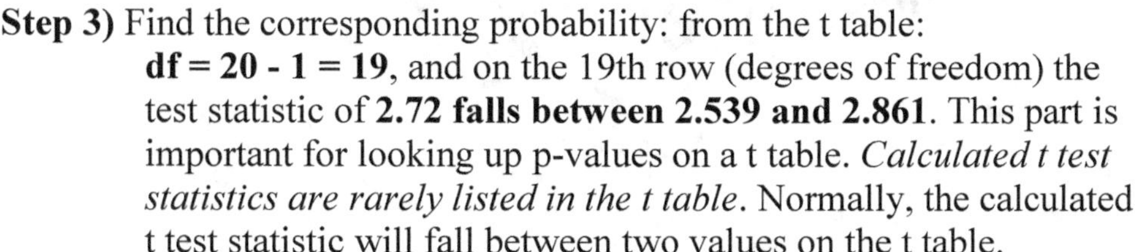

Step 3) Find the corresponding probability: from the t table:
df = 20 - 1 = 19, and on the 19th row (degrees of freedom) the test statistic of **2.72 falls between 2.539 and 2.861**. This part is important for looking up p-values on a t table. *Calculated t test statistics are rarely listed in the t table*. Normally, the calculated t test statistic will fall between two values on the t table.

Follow the two adjacent columns of 2.539 and 2.861 (either up or down, depending on which style of t table you have) until they intersect with the One-sided P (One-tail) row. The p-value of 2.72 falls between the two p-values from the One-sided P (one-tail) row, and those two values are .01 and .005.

(**Note: mgz recommends estimating the p-value with a real number for ease of use. Let's **pretend** that the p-value generated by a t statistic of 2.72 is **.008** because it is somewhere between .01 and .005... **.01 < .008 < .005**

t distribution critical values table

d.f.	Level of confidence, c	0.50	0.80	0.90	0.95	[illegible]	[illegible]
	One tail, α	0.25	0.10	0.05	0.025	0.01	0.005
	Two tails, α	0.50	0.20	0.10	0.05	[illegible]	[illegible]
1		1.000	[illegible]	[illegible]	[illegible]	31.82[illegible]	63.657
2		.816	[illegible]	[illegible]	4.303	6.9[illegible]	9.925
3		.765	1.638	2.353	3.182	[illegible]	5.841
4		.741	[illegible]	[illegible]	[illegible]	[illegible].747	4.604
5		.727	[illegible]	[illegible]	[illegible]	3.365	4.032
6		.718	1.440	1.943	2.447	3.143	3.707
7		.711	1.415	1.895	2.365	2.998	3.499
8		.706	[illegible]	[illegible]	2.306	2.896	3.355
9		.703	[illegible]	[illegible]	2.262	2.821	3.250
10		.700	[illegible]	[illegible]	2.228	2.764	3.169
11		.697	1.363	1.796	2.201	2.718	3.106
12		[illegible]	[illegible]	[illegible]	2.179	2.681	3.055
13		.694	[illegible]	1.771	2.160	2.650	3.012
14		.692	[illegible]	[illegible]	2.145	2.624	2.977
15		.691	[illegible]	1.753	[illegible]	2.602	2.947
16		.690	[illegible]	[illegible]	2.120	[illegible]	2.921
17		.689	1.333	1.740	2.110	[illegible]	2.898
18		.688	1.330	1.734	2.101	[illegible]	[illegible]
19		.688	1.328	1.729	2.093	2.539	2.861
20		.687	1.325	1.725	2.086	[illegible]	[illegible]
21		.686	1.323	1.721	2.080	2.518	2.831
22		.686	1.321	1.717	2.074	2.508	2.819
23		.685	1.319	1.714	2.069	2.500	2.807
24		.685	1.318	1.711	2.064	2.492	2.797
25		.684	1.316	1.708	2.060	2.485	2.787
26		.684	1.315	1.706	2.056	2.479	2.779
27		.684	1.314	1.703	2.052	2.473	2.771
28		.683	1.313	1.701	2.048	2.467	2.763
29		.683	1.311	1.699	2.045	2.462	2.756
∞		.674	1.282	1.645	1.960	2.326	2.576

The p-value of 2.72 falls between these two p-values

2.72 falls between these two critical values

Step 4) Compare the calculated p-value (**.008**) to the critical p-value (**.01**): .01 > .008... The critical value is greater than the calculated p-value. (***Remember the rules for null rejection?***)

Step 5) Either reject the null hypothesis, or not, based on the data: *Because the calculated p-value is* ***greater than*** *the critical value this data is indeed significant and we reject the null hypothesis.*

Step 6) What does this mean? *This study says that if we weighed twenty linemen from many different teams, then 99% of the mean weights of these linemen would be greater than 278 pounds.*

Two-sample t test statistic

The main difference between a one-sample and a two-sample t test is that a one-way t test compares a small sample to a stated mean, or norm, while a two-sample t test compares two samples (regardless of sample size) against each other to see if there is a significant difference between them.

Here is the formula for a ***two-sample t test statistic***:

$$t = \frac{\bar{X}_1 - \bar{X}_2}{\sqrt{\frac{s_1^2}{n_1} + \frac{s_2^2}{n_2}}}$$

Online video example 10.5: One 7-person group was given vitamin B-12 three times a day while another 7-person group was not given any vitamin supplements. After two weeks, both groups were asked to run 50 yards as fast as possible. Their times were recorded to the nearest second. **What is the t test statistic?**

Group A	Group B
15	18
14	19
16	17
12	16
14	14
12	14
10	18

Step 1) Make sure this a two-sample t test situation: we don't know the population mean or standard deviation, and the sample size is less than 30, so yes, this is a two-sample t test problem.

Step 2) Find the mean and standard error from both groups:

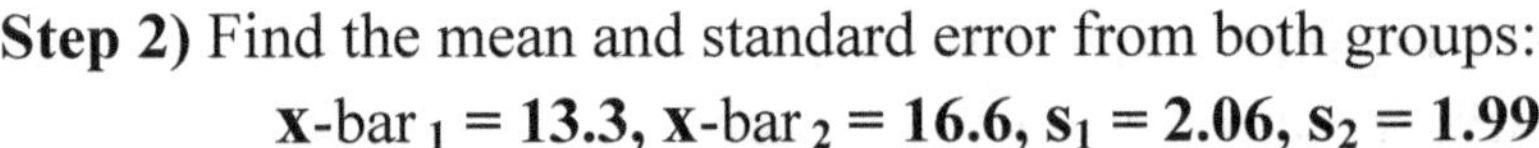
x-bar $_1$ = **13.3, x**-bar $_2$ = **16.6, s_1 = 2.06, s_2 = 1.99**

Step 3) Substitute the known values into the two-sample t test statistic formula:

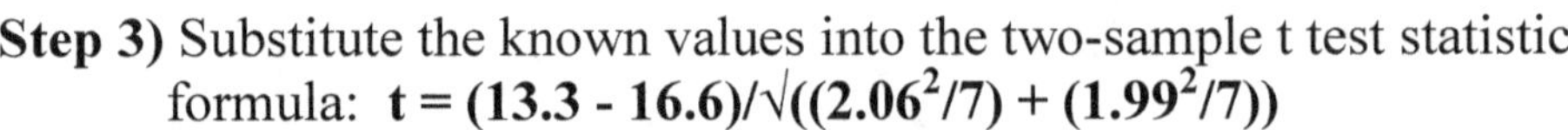
$t = (13.3 - 16.6)/\sqrt{((2.06^2/7) + (1.99^2/7))}$

Step 4) Math time, or should we say tea time?

$t = (13.3 - 16.6)/\sqrt{((2.06^2/7) + (1.99^2/7))} = -3.05$

Step 5) So the t test statistic from this problem is the ***absolute value of t***... |-3.05| = **3.05**.

Two-sample t test confidence interval

The confidence interval for a two-sample t test works just like every other confidence interval, but instead of using a single sample mean we use the difference between two sample means, plus/minus the margin of error.

*Note: If the two sample sizes are not the same size, ***use the smaller n*** to find the degrees of freedom.

t*

In the confidence interval formula for a t test, t* represents the critical value (from a table) corresponding to a level of confidence.

Here is the ***two-sample t test confidence interval*** formula:

$$(\bar{X}_1 - \bar{X}_2) \pm t^* \sqrt{\frac{s_1^2}{n_1} + \frac{s_2^2}{n_2}}$$

Online video example 10.6: One 7-person group was given vitamin B-12 three times a day while another 7-person group was not given any vitamin supplements. After two weeks, both groups were asked to run 50 yards as fast as possible. Their times were recorded to the nearest second. **Construct a 90% confidence interval for the mean.**

Group A	Group B
15	18
14	19
16	17
12	16
14	14
12	14
10	18

Step 1) Make sure this a two-sample t test situation: we don't know the population mean or standard deviation, and the sample size is less than 30, so yes, this is a two-sample t test problem.

Step 2) Find the mean and standard error from both groups:

x-bar $_1$ **= 13.3, x**-bar $_2$ **= 16.6,** s_1 **= 2.06,** s_2 **= 1.99**

Step 3) Find the t* critical value from the t table:
degrees of freedom = n - 1 = 7 -1 = 6. Now find where the 6th df row and the 90% confidence level C intersect. **t = 1.943**

t distribution critical values table

d.f.	Level of confidence, *c*	0.50	0.80	0.90	0.95	0.98	0.99
	One tail, *α*	0.25	0.10	0.05	0.025	0.01	0.005
	Two tails, *α*	0.50	0.20	0.10	0.05	0.02	0.01
1		1.000	3.078	6.314	12.706	31.821	63.657
2		.816	1.886	2.920	4.303	6.965	9.925
3		.765	1.638	2.353	3.182	4.541	5.841
4		.741	1.533	2.132	2.776	3.747	4.604
5		.727	1.476	[illegible]	2.571	3.365	4.032
6		.718	1.440	1.943	2.447	3.143	3.707
7		.711	1.415	[illegible]	2.365	2.998	3.499
8		.706	[illegible]	1.860	2.306	2.896	3.355
9		.703	[illegible]	1.833	2.262	2.821	3.250
10		.700	[illegible]	1.812	2.228	2.764	3.169
11			1.363	1.796	2.201	2.718	3.106
12			1.356	1.782	2.179	2.681	3.055
13			1.350	1.771	2.160	2.650	3.012
14			1.345	1.761	2.145	2.624	2.977
15			1.341	1.753	2.131	2.602	2.947
16		.690	1.337	1.746	2.120	2.583	2.921
17		.689	1.333	1.740	2.110	2.567	2.898
18		.688	1.330	1.734	2.101	2.552	2.878
19		.688	1.328	1.729	2.093	2.539	2.861
20		.687	1.325	1.725	2.086	2.528	2.845
21		.686	1.323	1.721	2.080	2.518	2.831
22		.686	1.321	1.717	2.074	2.508	2.819
23		.685	1.319	1.714	2.069	2.500	2.807
24		.685	1.318	1.711	2.064	2.492	2.797
25		.684	1.316	1.708	2.060	2.485	2.787
26		.684	1.315	1.706	2.056	2.479	2.779
27		.684	1.314	1.703	2.052	2.473	2.771
28		.683	1.313	1.701	2.048	2.467	2.763
29		.683	1.311	1.699	2.045	2.462	2.756
∞		.674	1.282	1.645	1.960	2.326	2.576

6 df and a 90% CI intersect here.

Step 4) Substitute the known values into the two-sample t test confidence interval formula:

$$\mathbf{(13.3 - 16.6) \pm 1.943(\sqrt{((2.06^2/7) + (1.99^2/7))}}$$

Step 5) Math time, oh boy!

$t = (13.3 - 16.6) \pm 1.943(\sqrt{((2.06^2/7) + (1.99^2/7))} =$
$(-3.3) \pm 1.943(\sqrt{(.61 + .56)} = (-3.3) \pm 1.943(1.08) =$
$\mathbf{-3.3 \pm 2.1 = [-5.4, -1.2]}$

Step 6) Interpret: *This interval means that if we repeated this experiment many many times then the difference between the means of the two groups would fall between 1.2 and 5.4 90% of the time.*

Two-sample t test -the entire process

Let's do a two-sample t test. Are you up for it? The steps are identical to a test of significance of a null hypothesis.

Online video example 10.7: One 7-person group was given vitamin B-12 three times a day while another 7-person group was not given any vitamin supplements. After two weeks, both groups were asked to run 50 yards as fast as possible. Their times were recorded to the nearest second. **Is there significant evidence at $\alpha = .05$ level that proves that vitamin B-12 make the people run faster? How about at the $\alpha = .01$ level?**

Group A	Group B
15	18
14	19
16	17
12	16
14	14
12	14
10	18

Step 1) Make sure this a two-sample t test situation: we don't know the population mean or standard deviation, and the sample size is less than 30, so yes, this is a two-sample t test problem.

Step 2) Find the mean and standard error from both groups:

x-bar $_1$ **= 13.3, x**-bar $_2$ **= 16.6,** $\mathbf{s_1 = 2.06}$, $\mathbf{s_2 = 1.99}$

Step 3) Substitute the known values into the two-sample t test statistic formula:

t = (13.3 - 16.6)/√((2.062/7) + (1.992/7))

Step 4) Math time, you lucky student!

t = (13.3 - 16.6)/√((2.062/7) + (1.992/7)) = **-3.05**

Step 5) So the t test statistic from this problem is the absolute value of -3.05 = **3.05**. (This is a huge test statistic so remember (again) to act surprised when we finish this problem, ok?)

Step 6) Find the p-value from the test statistic with a t table:

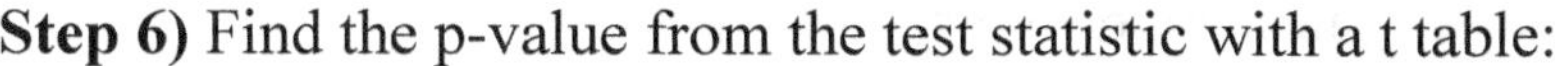

t distribution critical values table

d.f.	Level of confidence, c	0.50	0.80	0.90	[illegible]	[illegible]	0.99
	One tail, α	0.25	0.10	0.05	0.025	0.01	0.005
	Two tails, α	0.50	0.20	0.10	[illegible]	[illegible]	0.01
1		[illegible]	[illegible]	[illegible]	[illegible]	31.821	63.657
2		[illegible]	[illegible]	[illegible]	[illegible]	6.965	9.925
3		[illegible]	[illegible]	[illegible]	3.182	4.541	5.841
4		[illegible]	[illegible]	[illegible]	2.776	3.747	4.604
5		[illegible]	[illegible]	[illegible]	[illegible]	[illegible]	4.032
6		.718	1.440	1.943	2.447	3.143	3.707
7		.711	1.415	1.895	[illegible]	[illegible]	3.499
8		[illegible]	[illegible]	[illegible]	[illegible]	2.896	3.355
9		[illegible]	[illegible]	[illegible]	[illegible]	2.821	3.250
10		[illegible]	[illegible]	[illegible]	[illegible]	2.764	3.169
11		[illegible]	[illegible]	[illegible]	[illegible]	2.718	3.106
12		[illegible]	[illegible]	[illegible]	2.179	2.681	3.055
13		[illegible]	[illegible]	[illegible]	2.160	2.650	3.012
14		[illegible]	[illegible]	[illegible]	2.145	2.624	2.977
15		[illegible]	[illegible]	[illegible]	2.131	2.602	2.947
16		[illegible]	[illegible]	[illegible]	2.120	2.583	2.921
17		[illegible]	[illegible]	[illegible]	2.110	2.567	2.898
18		[illegible]	[illegible]	[illegible]	2.101	2.552	2.878
19		.688	1.328	1.729	2.093	2.539	2.861
20		.687	1.325	1.725	2.086	2.528	2.845
21		.686	1.323	1.721	2.080	2.518	2.831
22		.686	1.321	1.717	2.074	2.508	2.819
23		.685	1.319	1.714	2.069	2.500	2.807
24		.685	1.318	1.711	2.064	2.492	2.797
25		.684	1.316	1.708	2.060	2.485	2.787
26		.684	1.315	1.706	2.056	2.479	2.779
27		.684	1.314	1.703	2.052	2.473	2.771
28		.683	1.313	1.701	2.048	2.467	2.763
29		.683	1.311	1.699	2.045	2.462	2.756
∞		.674	1.282	1.645	1.960	2.326	2.576

The p-value of 3.05 falls between these two p-values

3.05 falls between these two critical values

Step 7) Compare the calculated p-value to the critical value:

Let's pretend our calculated p-value is .015 as an estimate because .01 < .015 < .025. We then compare the calculated p-value of .015 to an alpha of .05 and then compare it with an alpha of .01.

.05 > .015 is true, but .01 > .015 is not true.

Step 8) Make your decision and put this puppy to bed:

Since the critical value of .05 is greater than the calculated p-value of .015, then yes, this is a significant difference between the groups and we would **reject the null at α = .05.**

But, since the critical value of .01 is not greater than the calculated p-value of .015, then no, this is not a significant difference between the groups and we would **not reject the null at α = .01.**

In a nutshell, the data is significant at the .05 level, but not at the .01 level.

Chapter 10 Review

A ***t test*** of significance follows the same steps as a z test. Any study that meets at least one of the following three conditions should use a t test: 1) If the population standard deviation is unknown (***one-sample t test***); 2) If the sample size is thirty or less (***one-sample t test***); 3) If two samples from the same population (regardless of sample size) are being compared (***two-sample t test***).

The z and the t distribution formulas are identical, but their normal distribution tables are not. The z tables and the t tables are set up differently and have a different process to find specific probabilities.

Degrees of freedom are used to find p-values in a t distribution table. Remember is that the number of degrees of freedom is ***normally*** equal to ***one less that the sample size. df = n - 1***

The t distribution curve is similar to the z distribution curve in shape, but the ***fewer*** the degrees of freedom, the "shorter" and "wider" the curve becomes; t distributions have more area in their tails than a z curve. The more degrees of freedom, the more the t distribution approaches the standard normal distribution, the z distribution. In fact, ***anything greater than 29 degrees of freedom (n > 30) is to be treated as a z distribution***.

With t tests, we use the ***standard error*** which is the standard deviation divided by the square root of the sample size!

A ***t distribution table*** is used to find probabilities from t test statistics. The t table is different from a standardized normal distribution table in that it uses the number of degrees of freedom to calculate probabilities.

A ***one-sample t test,*** also known as a ***student t test,*** is used when the sample size is 30 or less, or when the population standard deviation is not known. A ***two-sample t test*** is used when two samples from the same population are compared.

Guided homework Chapter 10: t tests

1. A medical study of 21 people's blood pressure found that the mean was 114.9 with standard deviation of s = 9.3. What is the standard error?

2. What critical value t* from the t distribution table would you use for the CI for the mean population in each of the following situations?

   a) A 95% confidence interval based on 15 observations

   b) A 99% confidence interval based on 27 observations

   c) A 80% confidence interval based on 8 observations

3. The one sample t statistic from a sample of 25 observations from the two sided test of H_0: $\mu = 64$, H_a: $\mu \neq 64$ has the value of t test statistic = 1.02.

   a) How many degrees of freedom are there?

   b) Find the p-value from Table C

   c) Is the value of t test statistic = 1.02 statistically significant at alpha = 10% level?

   d) Is the value of t test = 1.02 statistically significant at alpha = 5% level?

4. You have a SRS of 15 observations from a normally distributed population. If the mean of the sample is 25.6 and the standard error is s = 2.35, what is a 98% confidence interval for the mean μ of the populations?

5. Here are measurements (in millimeters) of a critical dimension on a sample of automobile engine crankshafts:

224.121	223.213
223.461	223.982
224.098	223.976
224.008	223.909
224.089	223.989
224.057	223.902
223.517	223.761
223.987	223.581

The manufacturing process is known to vary normally with an unknown standard deviation. The process mean is supposed to be 224.000 mm. Do these data give evidence that the process mean is not equal to the target of 224 mm? State hypotheses and calculate a test statistic and its p-value. Are you convinced that the process mean is not 224 mm?

6. A study examined the learning of "Blissymbols" by children. Blissymbols are pictographs that are sometimes used to help learning-impaired children communicate. The researcher designed two computer lessons that taught the same content using the same examples. One lesson required the children to interact with the material, while in the other the children controlled only the pace of the lesson. Call these two styles "Active" and "Passive". After the lesson, the computer presented a quiz that asked the children to identify 56 Blissymbols. Here are the numbers of correct identifications by the 24 children in the Active group:

29 28 24 31 15 24 27 23 20 22 23 21

24 35 21 24 44 28 17 21 21 20 28 16

The 24 children in the Passive group had these counts of correct identifications:

16 14 17 15 26 17 12 25 21 20 18 21

20 16 18 15 26 15 13 17 21 19 15 12

Is there good evidence that active learning is superior to passive learning? State hypotheses, give a test and its p-value, and state your conclusion.

Chapter 10 Quiz

1. True or false?

A one-sample t test analyzes the difference between a sample group and a stated, or claimed, population average while a two-sample t test checks for a significant difference between two groups from the same population.

2. Amphibian Annie is testing the hypothesis that the average weight of the dreaded brown swamp toad is 157 grams. She catches 10 of the slimy fellows and gets a sample mean of 147 grams with a standard deviation of 16.6 grams.

a. What is the standard error?

b. Make a 95% confidence interval from this data for the true mean weight of the toads.

c. How many degrees of freedom are there?

d. What is the t test statistic?

e. What is the p-value?

f. Is the data significant at $\alpha = .05$? How about $\alpha = .01$?

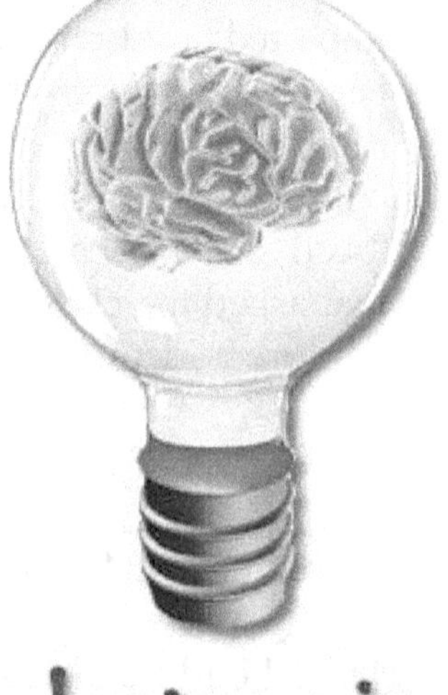

3. What would the p-value be from a t test statistic of 1.68 from a sample size of 27 on a two-tailed test?

4. True or false?

The p-values of a t test statistic are the same for a one-tailed and a two-tailed test.

5. True or false?

The standard deviation of a sample is called the standard error. (Assume that you will divide it by the square root of the sample size, yes?)

6. What conditions necessitate a t test?

7. A dog trainer claims that **he is better at** training dogs to catch more Frisbees than you. He takes six dogs and you take six dogs and work with them for a few hours and then put them to the test. The table lists the number of catches.

Trainer	You
22	26
21	26
22	27
28	25
27	31
22	30

a. Construct a 95% confidence interval from the data. (Assume it is a **not equal to** problem.)

b. What are the null and alternative hypotheses?

c. What is the t test statistic?

d. What is the p-value?

e. If this was a two-tailed test, would you reject the null at $\alpha = .05$?

f. If this was a two-tailed test, would you reject the null at $\alpha = .01$?

8. A t test is used to compare the variances between two groups. Which test would you use to compare three or more groups?

a. drug test
b. test of means
c. binomial test
d. z test
e. ANOVA

9. True or false?

The number of degrees of freedom is equal to the sample size.

10. An alligator-ologist is claiming that the stated average length of one-year old alligators ***is not*** 1.66 meters as stated in the scientific journal of Alligators Weekly. He did his own study and came up with the following lengths in meters.

1.65	**1.44**
1.55	**1.32**
1.34	**1.71**
1.74	**1.56**
1.81	**1.42**
1.39	**1.8**
1.51	**1.62**

a. What are the null and alternative hypotheses?

b. What is the test statistic?

c. What is the p-value?

d. Is the data significant at $\alpha = .05$? How about $\alpha = .01$?

Chapter 11: Sample Proportions (p-hat)

Chapter

11

sample proportions

($\hat{p}$)

4 out of 5 dentists agree...

Up to this chapter, we have been making inferences from sample means about population means. But now we are going to learn how to make inferences from ***proportions*** of populations. The steps to solving these new proportion inferences are practically identical to what we have already learned, but instead of sample means we will use proportions from populations.

In this chapter, we will investigate inferences using proportions:

- Sample proportions (***p-hat***)

- The mean and standard deviation of population proportion **p**

- Conditions for inference about **p**

- Confidence intervals for ***p-hat***

- Margin of error and sample size for ***p-hat***

- Tests of significance for ***p-hat***

- Sample proportions ($\hat{p}$)
- The mean and standard deviation of p
- Conditions for inference about p
- Confidence intervals for $\hat{p}$
- Margin of error and sample size for $\hat{p}$
- Test of significance for $\hat{p}$

Parameters or Statistics

Parameters are data that is true about a **POPULATION** and are represented with Greek letters.

Statistics are data that is true about a **SAMPLE** from a **POPULATION** and are represented with English letters.

Sample proportions

A ***proportion*** is a part of a population. Example: 4 out of 28 of Ms. Gupta's algebra students are left-handed. The proportion for left-handedness is 4/28 for her algebra class. How accurate is that proportion compared to all people everywhere?

A ***sample*** is also a part of a population but we will not be finding the means of these samples. Rather, we will use proportions to make inferences about populations. We use the same steps that we used when making inferences about population means from sample means. The formulas are practically the same.

In the first part of this chapter we will learn to estimate a **parameter** (something true about a population) using the data (**statistics**) gathered from a ***sample proportion***.

We will be ***estimating*** the *unknown* proportion of a population. We will call the specific outcome that we are trying to estimate a "success" like we do using binomial probabilities. In our example above, a success would be a left-handed person.

With proportions, ***p-hat*** estimates the population's parameter **p**.

Important: a single **p represents the **stated population proportion**. Below is the formula for the sample proportion ***p-hat***.

$$\hat{p} = \frac{\textbf{number of successes (in the sample)}}{\textbf{total number of individuals in the sample}}$$

Example 11.1: Ms. Tran had 38 students in her 3rd period geometry class. Twenty five of the students were Hispanic. What is the sample proportion (***p-hat***) for Hispanic students?

Step 1) Break out the formula above and substitute known values.
# of success = # of Hispanic students = 25, out of 38 total students.

$$\hat{p} = \frac{25}{38} \qquad \hat{p} = \frac{X}{n}$$

number of successes

total number (sample size)

The stated population proportion (the claim) is represented as a plain p.

Step 2) Change your fraction into a decimal: 25/38 $\approx$ 0.658

Step 3) So, the ***sample proportion*** (***p-hat***) (of Hispanic students) is **0.658**. We call the sample proportion *p-hat* because with proportions, it's always a party!

Sampling distribution of $\hat{p}$

How accurate is *p-hat* as an estimate of parameter **p**? How did we check the accuracy when we were calculating with sample means? We took many different samples at random. We do the same thing here. We need to take many different proportions at random from a population. And if the sample size is large then the many different proportions from a population should form a normal distribution.

Here are the rules:

Sampling distribution of $\hat{p}$

From an SRS of size n from a large population that has a proportion p of success, then ...

$$\hat{p} = \frac{\text{number of successes in the sample}}{n}$$

1) As the sample size increases, the sampling distribution $\hat{p}$ approaches a normal distribution.

2) The ***mean*** of the sampling distribution is p (the population's proportion of successes)

3) The ***standard deviation*** of the sampling distribution is

$$\sqrt{\frac{p(1-p)}{n}}$$

Vocabulary

SRS

means

Sample

Random

Space.

SRS

also stands for

Simple

Random

Sample.

An SRS is a randomly chosen sample from a population.

Here is the graph of the normal distribution of ***p-hat***:

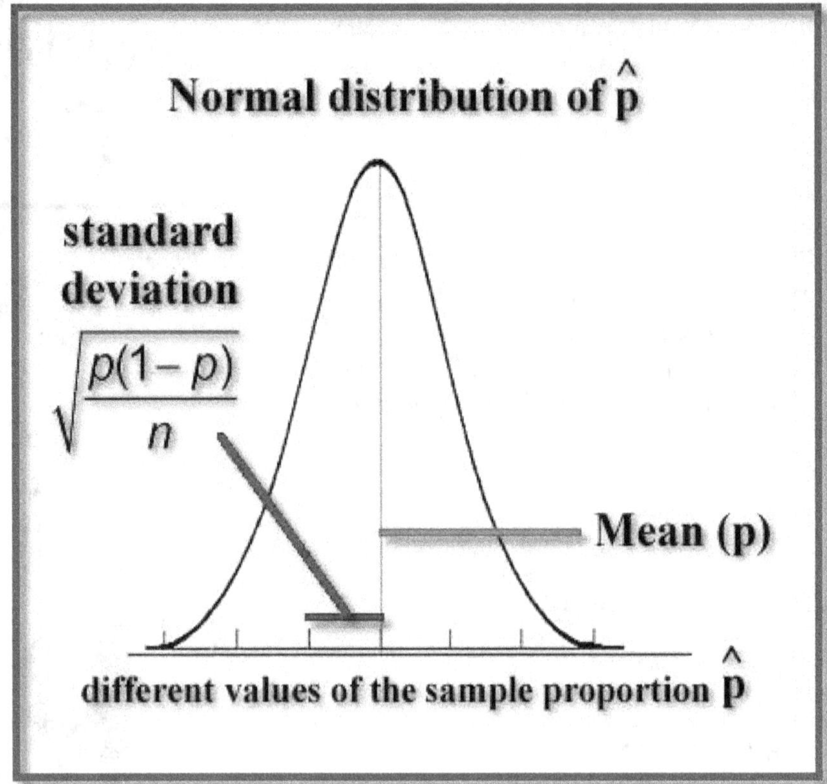

Remember the ***Law of Large Numbers***? It states that the larger the sample size the more accurate the statistics that describe the population's parameters. This rule also applies to proportions. In fact, if the sample size is not large (compared to the population) then we should not use the proportion formulas.

The cut-off number for sample size changes from problem to problem, but a good rule of thumb is if the population is large then the sample size should be a small percentage, but if the population is small then the sample size should be a large percentage. Mgz normally sticks to the "rule of 30" from the standardized (z) testing.

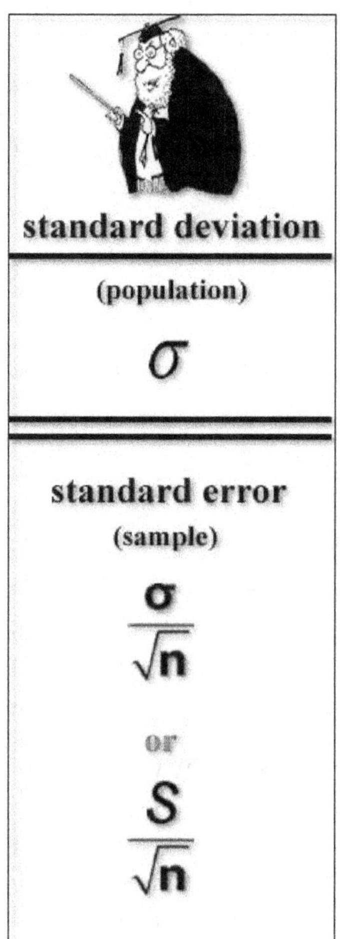

And we must not forget about the ***Central Limit Theory***. One of its rules states that the larger the sample size then the smaller the standard deviation. Remember why? That's correct – we have to **divide** the population **standard deviation** by the **square root** of the sample size **n**. So, when n increases, the value of the standard deviation decreases. When we divide the standard deviation by the square root of the sample size we can call that the ***standard error***, but you already knew that by now, yes?

*Important note: some textbooks and professors frequently refer to a standard error as a standard deviation. The two are not the same thing but are very closely related. The general concept is the same in that they both relate the amount of variance in the data. Remember that a sample's standard deviation has to be divided by the square root of n and that makes it a standard error.

We use the same rules to standardize ***p-hat*** that we use to standardize sample means.

You should already know this standardization formula from finding z scores from sample means (Chapter 7):

standardized sample means

$$z = \frac{\bar{X} - \mu}{\sigma / \sqrt{n}}$$

Here is the standardization formula for sample proportions (***p-hat***).

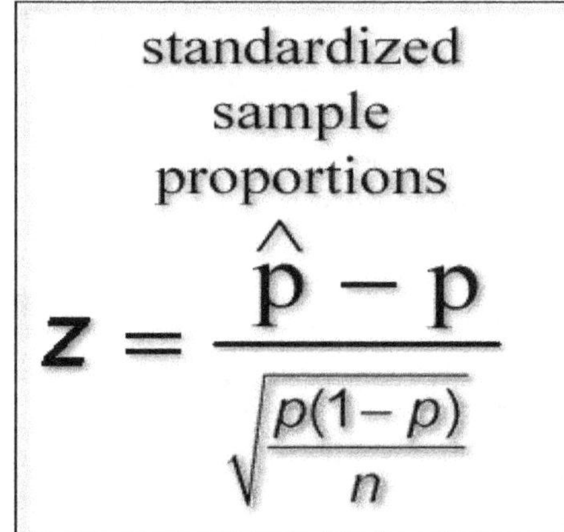

Making inferences about a true population proportion from a sample proportion follows the same set of rules that we use to make inferences about population means from sample means. Remember the conditions that must be true: most importantly, the proportions have to be ***randomly*** chosen, and the sample size has to be large enough that the distribution of *p-hat* is **normal**.

Okay, let's do one of these problems to cement the process and the interpretation.

Example 11.2: California community colleges claim that only 10% of their incoming freshmen are prepared for their more advanced algebra courses. If you selected 1,000 incoming freshman at random, what is the probability that **at least** 12% of them will be algebra proficient?

Step 1) Decipher the problem. We need the following information: what is **p**? What is ***p-hat***? What is ***n***? What is the sample's ***standard error***? How do we standardize the data? And most importantly, what exactly is the question asking?

Step 2) What is **p**? Remember that **p** represents the **population's** proportion. In this problem, we will use the community colleges' statement that 10% of incoming freshman are adequately prepared for their algebra courses. So, 10% is the population's proportion, or **p**.

p = 10% = 0.1

What is *p-hat*? ***Remember that p-hat has to be taken from a sample!*** This is how to tell the difference between **p** and ***p-hat***. So *p-hat* has to be the 12% (the sample proportion).

***p-hat* = 12% = 0.12**

What is ***n***? ***n* = 1,000**

What is the **sample's standard error**? Break out the standard deviation formula for proportions and then plug and chug.

$$\sqrt{\frac{p(1-p)}{n}} \Rightarrow \sqrt{\frac{.1(.9)}{1000}} \Rightarrow .0095$$

Step 3) Find the **z score** (the standardized score) with the formula.

$$z = \frac{\hat{p} - p}{\sqrt{\frac{p(1-p)}{n}}}$$

Substitute known values:

$$z = \frac{.12 - .1}{.0095}$$

We get a final z score of 2.11: **z = 2.11**

Step 4) What exactly is the question asking? What we want to find out is the probability that **at least** 12% of these 1,000 randomly chosen students will be ready for college algebra using the fact that only 10% of the incoming freshmen are at the proficient level or higher. So, we are looking for the probability (**capital P**, or p-value) that the sample proportion will be equal to or greater than 12%.

P(*p-hat* ≥ .12) (we already found the z score from step 3)

… which goes to … **P(z ≥ 2.11)** … which goes to …

1 - .9826 (we subtract the probability value from the z score of 2.11 (.9826 from the z table) from 1 because it is a <u>**greater than**</u> problem) … which goes to … **.0174.**

And that means (finally) that if you repeated taking random samples of 1,000 incoming freshmen that less than 2% of the samples will contain at least 12% of the students who are ready for college algebra. And mgz can testify to it too!

z*

In the confidence interval formula for a z test, z* represents the critical value (from a table) corresponding to a a level of confidence.

Estimating population proportion p with a confidence interval

We are going to make estimations of true population proportions (**p**) from sample proportions (***p-hat***). Sound familiar? It should. We will follow the same steps as we did trying to estimate the true population mean from a sample mean.

Remember the confidence interval for a sample mean?

Estimate ± margin of error

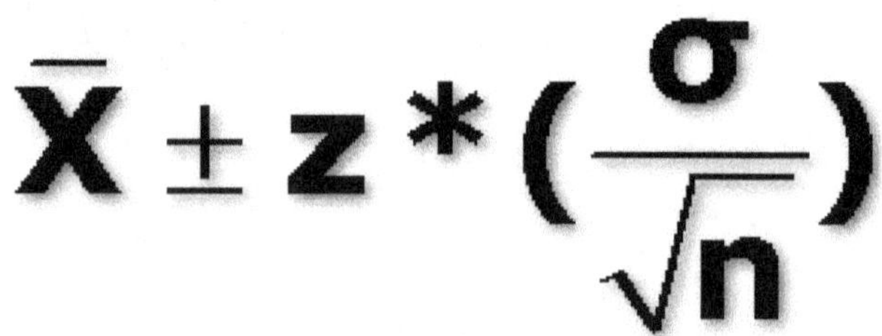

Well, here is the formula for a proportion confidence interval:

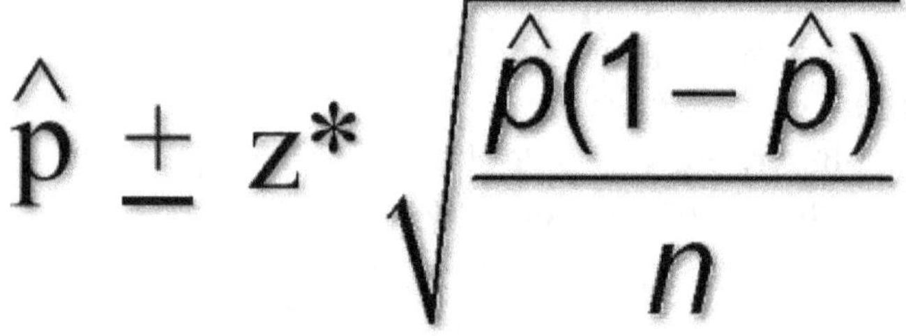

margin of error for a proportion

$$m = \pm z^* \sqrt{\frac{\hat{p}(1-\hat{p})}{n}}$$

(z* is the critical value corresponding to the confidence level C)

This confidence interval is still the **estimate ± margin of error**. The estimate of the true population proportion is the sample proportion (*p-hat*) and the margin of error is made up of the product between the specified critical value (z* - depending on the level of confidence C) and the sample proportion's standard error.

Notice that we do ***not*** know what the population proportion (**p**) is, so we use the sample proportion's (***p-hat***) standard error.

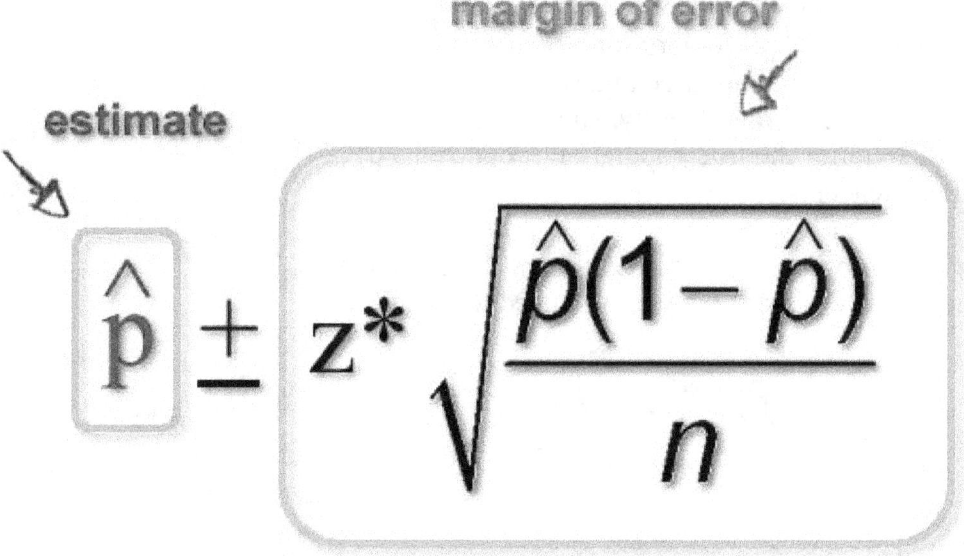

In the graphic below, we are confident that ***p-hat*** lies within the margin of error of the unknown population proportion **p**.

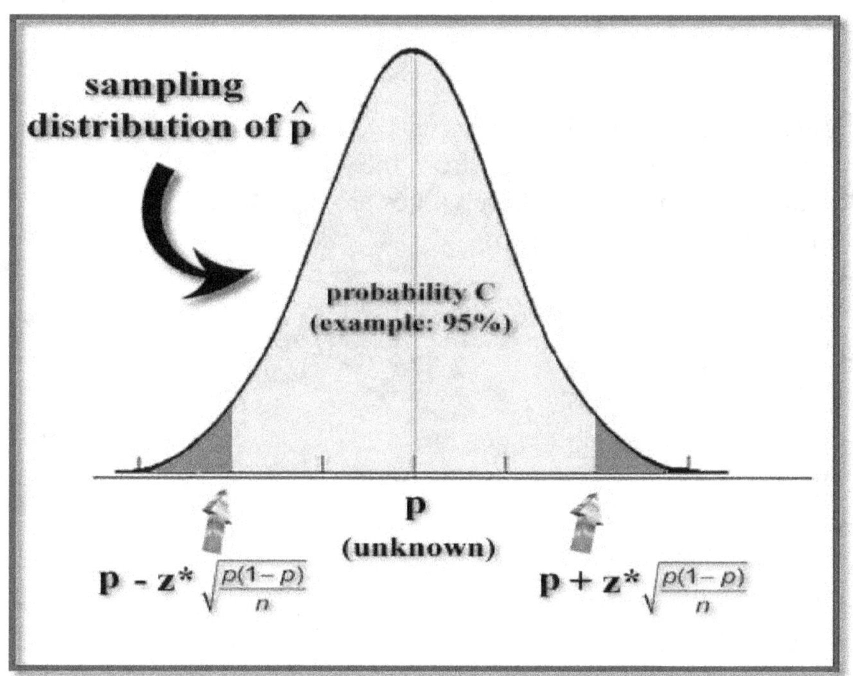

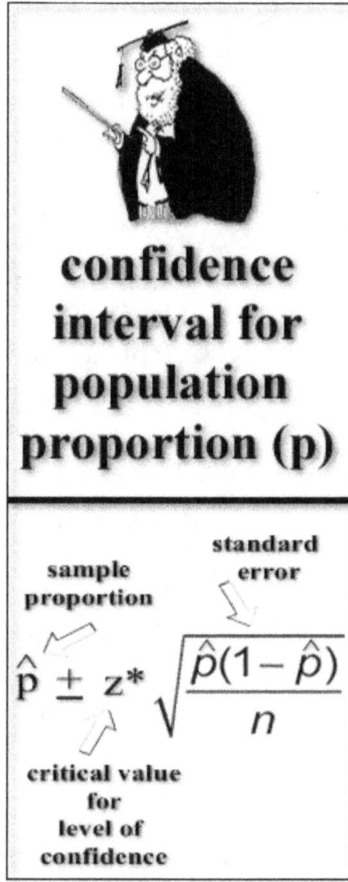

Example 11.3: In 1960, 75 percent of college teachers were full-time tenured or tenure-track professors, according to the New York Times. Let's pretend that you are doing a current study on the percentage of adjunct college teachers and randomly contacted 378 professors and asked them if they were adjunct faculty. Two hundred and seventy six said that they were indeed adjunct faculty. Build a 99% confidence interval for the true population proportion (**p**) of adjunct college teachers.

Step 1) Break out the sample proportion confidence interval formula and decide which pieces of data that you will need to substitute.

$$\hat{p} \pm z^* \sqrt{\frac{\hat{p}(1-\hat{p})}{n}}$$

Step 2) What is ***p-hat***? What is **z*** and what is **n**?

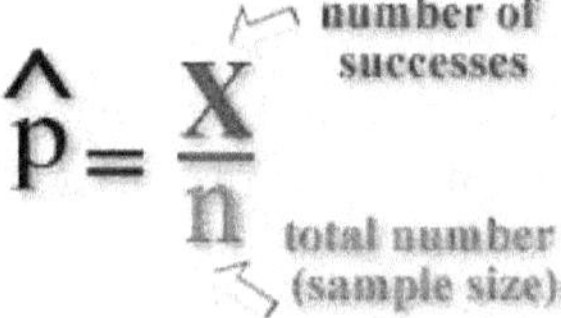

p-hat = X/n … (**X** is the number of successes, ***n*** is the total number in the sample) … *p-hat* = **276/378 …** ***p-hat*** **≈ 0.73**

z* is the critical value corresponding to the level of confidence.

Remember this from previously covered confidence intervals? The level of confidence C tells us which critical value (z*) from the table to use when we build our margin of error. (These values are from the normal distribution (z) table.)

confidence level C	**90%**	**95%**	**99%**
critical value z*	**1.645**	**1.960**	**2.576**

So, we will use 2.576 because that is the critical value corresponding to the 99% CI.

z* = 2.576

n = 378

Step 3) Substitute the known values.

$$\hat{p} \pm z^* \sqrt{\frac{\hat{p}(1-\hat{p})}{n}} \quad \text{... goes to ...} \quad .73 \pm 2.576 \sqrt{\frac{.73\,(1 - .73)}{378}}$$

... which finally goes to ... **.73 ± .06**

Step 4) Interpret the confidence interval.

If this study was conducted repeatedly using groups of 378 randomly chosen college teachers, then 99% of the sample proportions should have roughly between 67% and 79% adjunct faculty, and the true population proportion (p) of all adjunct college teachers should also fall within that range.

Roughly 3 out of 4 college professors are adjunct faculty.

So, in the last 50 years the percent of full-time college teachers has fallen from 75% to around 27%. It's usually about the money.

Tests of significance for a proportion

How do we tell if one specific sample proportion reflects the true population proportion? Well, we use the same standardized test statistic process as we did before with testing a sample mean against a population mean.

The null hypothesis for a proportion is $\mathbf{H_0: p = p_0}$

Here is the formula to find the z test statistic for a proportion:

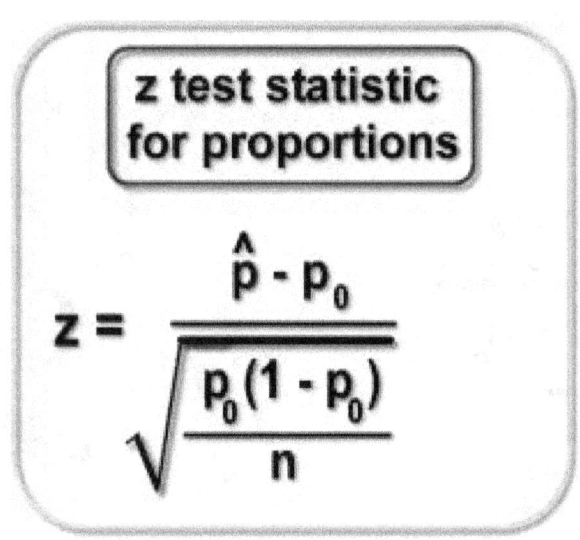

The approximate probability (**P**) values for standardized proportion tests against the H_0 are the p-values from a z score. (The ***p*** represents the true population proportion and the p_0 represents the proportion from the null hypothesis; this is the claim that is being tested.)

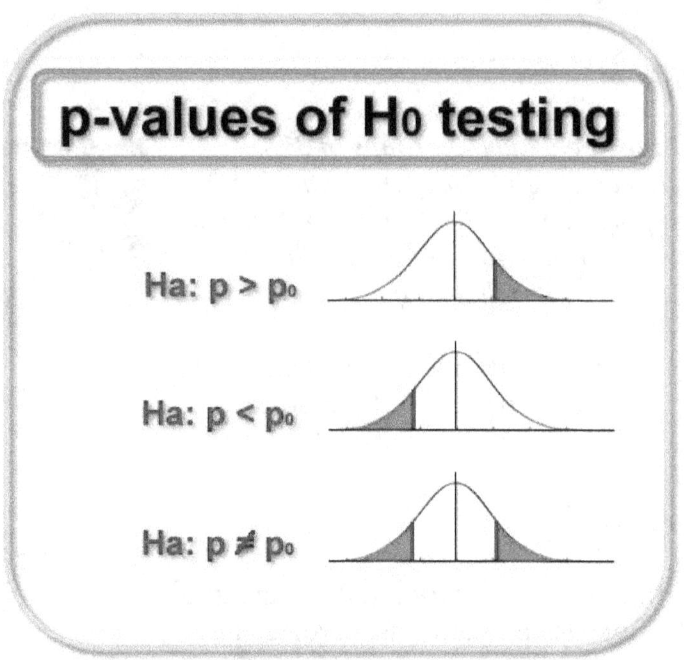

Example 11.4: Navy re-enlistment rates are at an all time high, with 63 percent of first-term sailors signing up for additional service. Let's say that you have contacted 82 first-term sailors and 51 of them planned on re-enlisting. Is this sample proportion good evidence that the stated re-enlistment proportion is accurate?

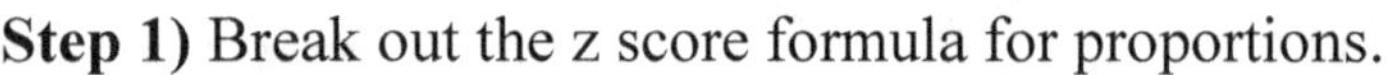

Step 1) Break out the z score formula for proportions.

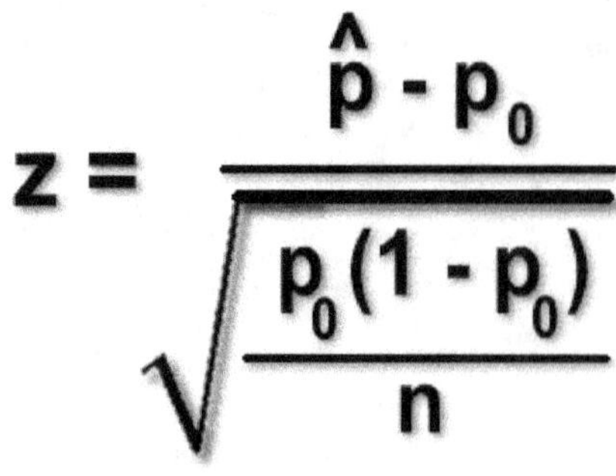

$$z = \frac{\hat{p} - p_0}{\sqrt{\frac{p_0(1 - p_0)}{n}}}$$

Step 2) Find ***p-hat***. ***p-hat*** = **51/82** ≈ **.62**

Step 3) Substitute known values to find the z score.

$$z = \frac{\hat{p} - p_0}{\sqrt{\frac{p_0(1 - p_0)}{n}}} \quad \text{... goes to ...} \quad z = \frac{.62 - .63}{\sqrt{\frac{.63(1 - .63)}{82}}}$$

... goes to ... **z ≈ -.01/.05 = -0.2**

Step 4) Look up the corresponding **p-value** for a z score of **-0.2**.
(Assume this is a less than problem.)

P(z < -0.2) = 0.4207

Step 5) Interpret.

Since the probability value is <u>not</u> less than 0.05, then this sample proportion is a good example that supports the claim that 63% of first-time sailors re-enlist.

Sample size for a proportion

The number of individuals in a sample proportion is always important. How do we know if we have a big enough ***sample size***? Well, there is a formula to find out. There is always a formula, yes?

On the next page is the formula to find the proper sample size for a sample proportion. (The z^* represents the critical value, like always, corresponding to the level of confidence; the ***m*** is the margin of error (the **± amount**), and **p*** is normally the population proportion, or if that is unknown then we use the sample proportion.)

$$n = \left(\frac{z^*}{m}\right)^2 p^*(1 - p^*)$$

Example 11.5: Navy re-enlistment rates are at an all time high, with 63 percent of first-term sailors signing up for additional service. How many first-term sailors should you contact if you wanted to build a 90% confidence interval with a margin of error of $\pm$ 0.05?

Step 1) Break out the sample size formula and substitute known values.

$$n = \left(\frac{z^*}{m}\right)^2 p(1 - p) \text{ ... goes to ... } n = \left(\frac{1.645}{.05}\right)^2 .63(1 - .63)$$

... goes to ... $n = 252.3$ and we always round up, so **n = 253.**

Comparing two proportions

When we want to compare something about two different populations we use independent proportions from each of the populations and compare them to investigate if there is a ***significant*** difference.

Since there are two groups, we need to label the data to identify from which group it belongs.

population #1		population #2	
population proportion	p_1	population proportion	p_2
sample proportion	$\hat{p}_1$	sample proportion	$\hat{p}_2$
sample size	n_1	sample size	n_2

(The subscripted number tells us which population the data is from.)

Dealing with two different sample proportions from two separate populations, we need to "**pool**" the information from both samples and populations.

Pooled sample proportion

$$\hat{p} = \frac{\text{the combined number of successes from both samples}}{\text{the combined totals of individuals in both populations}}$$

$$\hat{p} = \frac{X_1 + X_2}{n_1 + n_2}$$

Looking for significance between two populations is simply checking the **difference** between the two stated population parameters: $\mathbf{p_1 - p_2}$.

Here is the formula for **standard deviations** for **population** proportions:

Population proportions standard deviation

(when comparing two populations)

$$\sigma = \sqrt{\frac{p_1(1 - p_1)}{n_1} + \frac{p_2(1 - p_2)}{n_2}}$$

Looking for significance between two sample proportions is checking the difference between the two sample proportions: ***(p-hat$_1$) – (p-hat$_2$)***.

Here is the formula for the **standard error** for **sample** proportions:

Sample proportions standard error

(when comparing two sample proportions)

$$S = \sqrt{\frac{\hat{p}_1(1-\hat{p}_1)}{n_1} + \frac{\hat{p}_2(1-\hat{p}_2)}{n_2}}$$

Here is the confidence interval formula for the population proportions:

z*

In the confidence interval formula for a z test, z* represents the critical value (from a table) corresponding to a a level of confidence.

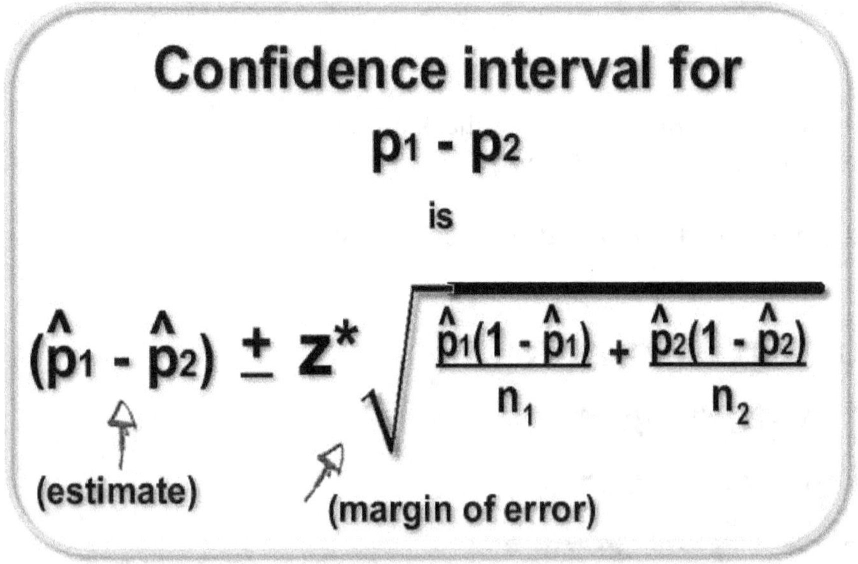

And finally here is the formula for the z score test statistics used to test the null hypothesis of $\mathbf{H_0: p_1 = p_2}$.

z score formula
(To test $H_0: p_1 = p_2$)

$$z = \frac{\hat{p}_1 - \hat{p}_2}{\sqrt{\hat{p}(1 - \hat{p})\left(\frac{1}{n_1} + \frac{1}{n_2}\right)}}$$

($\hat{p}$ in these tests is the pooled sample proportion.)

$$\hat{p} = \frac{X_1 + X_2}{n_1 + n_2}$$

Example 11.6: A study at a large university wanted to see if gender had anything to do with their students using their free tutoring services. Out of 846 female students, 543 said that they had used a tutor at least three times in the previous semester, while 274 male students out of 521 said that they used a tutor. A) Make a 95% confidence interval for the difference between the genders that used the tutoring services.
B) Proportionally speaking, is there a significant difference between female college students seeking out tutoring and their male counterparts?

Step 1A) Let's label the girls as $\mathbf{p_1}$ and the boys as $\mathbf{p_2}$.

Step 2A) Which pieces of data do we need to in order to solve this problem? Here is the appropriate confidence interval formula:

$$(\hat{p}_1 - \hat{p}_2) \pm z^* \sqrt{\frac{\hat{p}_1(1 - \hat{p}_1)}{n_1} + \frac{\hat{p}_2(1 - \hat{p}_2)}{n_2}}$$

Step 3A) Solve for the two ***p-hats***:

Girls: *p-hats*$_1$ = **543/846 ≈ .64**
Boys: *p-hats*$_2$ = **274/521 ≈ .53**

Step 4A) Substitute known values into the formula:

$$(.64 - .53) \pm 1.960 \sqrt{\frac{.64(1 - .64)}{846} + \frac{.53(1 - .53)}{521}}$$

Step 5A) Math time:

.11 ± .054

Step 6A) Interpret: *If this study was repeated many times under the same conditions, then the differences between the two population proportions would fall within this range **95%** of the time:* ***[.0546, .1154].***

Step 1B) Are there significantly more female students using the tutoring services? We are testing this null hypothesis: $\mathbf{H_0: p_1 = p_2}$.

We must find the z score from the data. Here is the formula:

$$z = \frac{\hat{p}_1 - \hat{p}_2}{\sqrt{\hat{p}(1 - \hat{p})\left(\frac{1}{n_1} + \frac{1}{n_2}\right)}}$$

Step 2B) ***<u>First we must find the pooled sample proportion</u>***.

$$\hat{p} = \frac{X_1 + X_2}{n_1 + n_2}$$

$$\text{p-hat} = \frac{543+274}{846+521} = \frac{817}{1367} \approx .5977 \approx .60$$

Step 3B) Substitute known values and solve for the z test statistic.

$$z = \frac{.64 - .53}{\sqrt{.6(1 - .6)\left(\frac{1}{846} + \frac{1}{521}\right)}}$$

z = 4.07

Step 4B) Look up the p-value for a z score of 4.07. (Use the ≠ method.)

2P(z > 4.07) = 2(1 - .9999) = .0002

Step 5B) Interpret: *Since our calculated p-value is less than .05 we reject the null hypothesis (which always states that there is no significant difference between the groups) and have statistically proven that there is a significant difference between the proportions of males and females using the tutoring services.*

Chapter 11 Review

A ***population's proportion*** is a claim about a population and is represented by the lone letter "**p**".

A ***sample proportion*** is a randomly selected sample from a population with **X** number of successes out of the total number of the sample size ***n***, and is represented by ***p-hat***. ***p-hat*** = **X/n**

$$\hat{p} = \frac{X}{n}$$

number of successes

total number (sample size)

When **n** is large enough (mgz says that 30 or larger is good) then the different sample proportions (*p-hat*) will have a normal distribution and we assume that *p-hat* = **p** with a standard deviation of $\sqrt{\frac{p(1-p)}{n}}$.

The ***confidence interval*** formula for the true population **p** from a sample proportion ***p-hat*** with a level of confidence C (z*) is:

margin of error

estimate

$$\hat{p} \pm z^* \sqrt{\frac{\hat{p}(1-\hat{p})}{n}}$$

The ***sample size*** needed to build a confidence interval with the approximate margin of error **m** for a population proportion **p** is:

$$n = \left(\frac{z^*}{m}\right)^2 p^*(1 - p^*)$$

(Use the population proportion p if known, if not use the sample proportion $\hat{p}$.)

A ***significance test*** for the null hypothesis of $\mathbf{H_0: p = p_0}$ is done using the standardized (z) test. Here is the formula:

$$z = \frac{\hat{p} - p_0}{\sqrt{\frac{p_0(1 - p_0)}{n}}}$$

Probability values (p-values) from the calculated z score are looked up in the Normal Distribution (z) tables.

Here is the formula to compare two population proportions using a z score and pooled sample proportion:

z score formula
(To test H_0: $p_1 = p_2$)

$$z = \frac{\hat{p}_1 - \hat{p}_2}{\sqrt{\hat{p}(1-\hat{p})\left(\frac{1}{n_1} + \frac{1}{n_2}\right)}}$$

($\hat{p}$ in these tests is the pooled sample proportion.)

$$\hat{p} = \frac{X_1 + X_2}{n_1 + n_2}$$

Guided homework Chapter 11: Sample Proportions

1. Left-handedness is a recessive genetic trait. If H is the gene for right-handedness and h is the gene for left-handedness, then all people have one of the following gene combinations from their parents: HH, Hh, hH or hh. Only the people with the hh genes will be left-handed, or supposedly around 25% of all people on earth should be left-handed, but most studies report that a more realistic proportion of left-handed people is around 12%.

a) Suppose the 12% proportion is correct. If you took many random samples of 72 people at a time, what should the **mean** of the sample proportions be close to? In other words, how many lefties out of the 72 people should there be?

b) What is *p-hat*?

c) What is the sample proportion's standard deviation (standard error)?

d) Build a 95% confidence interval for the true proportion of left-handedness for repeated samples of 72 people.

e) How many people should be surveyed if you want a margin of error of .01 and a confidence interval of 90%?

f) Any suggestions as to why only around 12% of people are left-handed when the mathematics tells us that around 25% of us should be left-handed?

2. In a 2009 survey, 821 college students were asked if they owned a cell phone that could connect to the internet and 534 of the respondents said that they did own a phone that could connect them to the Internet.

a) What is the sample proportion (*p-hat*)?

b) What is the standard error of the sample proportion?

c) What is the margin of error for a 95% confidence interval?

d) Build a 95% confidence interval for the proportion of students who have a cell phone that can connect to the internet.

3. A survey done in 2001 reported that 58% of American households watched more than 7 hours of television a day. You decide to test this proportion and see if it applies to your own neighborhood. In a random sample of 100 from your neighborhood households, 62% reported back that they did watch more than 7 hours of television a day.

a) What are the null and alternative hypotheses?

b) Is this proof that your neighborhood differs significantly from the 58% from the 2001 survey?

4. Remember the Pepsi Challenge? It was a blind taste test between Pepsi and Coca Cola. The Pepsi guys did the study and they had 50 people take the test with 29 of the tasters saying that they preferred Pepsi over Coke. Is this enough evidence to prove that most people preferred Pepsi?

5. Federal Work Study (FWS) is a great way for college students to work off some of their student loans, but not everyone takes advantage of this program. A SRS sample survey was done and 42 out of 63 female students said that they were working at their school earning FWS, and 25 out of 35 male students said that they were earning FWS.

a) What would the null and alternative hypotheses be if we were testing that more males earn FWS than females?

b) What are the sample proportions for the males and the females?

c) What is the standard error?

d) Make a 99% confidence interval for the difference between the sample proportions of FWS students.

e) What is the pooled sample proportion for all students receiving FWS?

f) Is there a significant difference between the proportions of male and female FWS students?

Chapter 11 Quiz

1. The Humane Society states that close to 39% of US households have at least one dog. Jimmy, a curious young statistician, does his own study and checks out 125 random households and finds out that 56 of them do have at least one dog.

a) What is the sample proportion from Jimmy's study?

b) If Jimmy repeated this same study one thousand times, what should the average number of dog-owner households from the sample proportions of 125 be?

c) What is the sample proportion's standard error?

d) Build a 95% confidence interval for the true proportion of dog-owners for repeated samples of 125 households.

e) How many people should be surveyed if you want a margin of error of .05 and a confidence interval of 99%?

2. The Department of Health reports the percentages of the population blood types in the table on the right. A local blood bank collects 67 pints in one day, of which 9 were A negative.

O	Positive	38.40%
A	Positive	32.30%
B	Positive	9.40%
O	Negative	7.70%
A	Negative	6.50%
AB	Positive	3.20%
B	Negative	1.70%

a) What is the sample proportion (*p-hat*) for the A negative blood donors out of the 67 donors?

b) What is the standard error of the sample proportion for the A negative blood donors from the table? (use the population **p = .065**)

c) What is the margin of error for a 95% confidence interval for the sample proportion of A negative blood donors?

d) Build a 95% confidence interval for the proportion of A negative blood donors.

e) Does this sample proportion support the data from the Department of Health about the percentage the population who have A negative blood?

3. A survey done in 2006 reported that 55% of American taxpayers supported a new proposal that would allow government funds to be used to send students from underperforming schools to a private school. A year later a for-profit online school performed their own survey and stated that 535 out of 961 randomly selected taxpayers replied that they would support sending students to private schools with government funds. The for-profit school says that this is evidence that more than 55% of taxpayers support this proposal.

a) What are the null and alternative hypotheses?

b) Is this significant proof that more than 55% of taxpayers support the proposal?

4. A new radio station is trying to decide which type of musical format to offer its listeners. Out of a poll of 1216 people who called in, 653 of them voted for classic rock and roll. Is this enough evidence to prove that most people preferred classic rock and roll? Is this a valid study?

5. Tattoos have become popular. A trendy fashion magazine randomly polled 1121 males between 18 to 35 years of age and 654 said that they had at least one tattoo, while 442 out of 657 females from the same age group said that they had at least one tattoo. According to this data, is there a significant difference between the proportions of the male and female tattooed people?

Chapter 12: Chi-square tests

chi-square

χ^2

chi-square

Chapter 12

chi-square tests

The type of data that a variable "produces" identifies which type of variable (quantitative or categorical) we are working with, which in turn helps us decide which type of statistical test to use for our data analysis.

A categorical variable is a variable that places the individual into a specific category. Examples: What is your gender? Are you left-handed or right-handed? What is your blood type? Notice that a single individual can be in more than one category. From the example, a left-handed female with blood type O positive would be in three different categories. But, how do we find out if a categorical variable is "behaving" like a population parameter?

In this chapter we will investigate whether distributions of categorical variables differ from one another using a ***chi-square test***.

What's in this chapter

- Two-way tables
- Expected cell counts
- Chi-square tests
- Goodness of fit
- Test for independence

Two-way tables

Since categorical variables put individuals into specific categories, the representation of the data takes on a different format using **rows** and **columns** representing the separate categorical variables. Categorical variables are also known as **nominal** variables.

In the following example, one categorical variable is grade level and another categorical variable is the type of dessert that a student prefers. Please note that **each subject is in two categories**, thus the name of ***two-way tables***.

two-way table (dessert preferences)

(grade level)	cupcake	ice cream	candy bar
elementary	336	252	153
middle school	77	91	68
high school	33	54	162

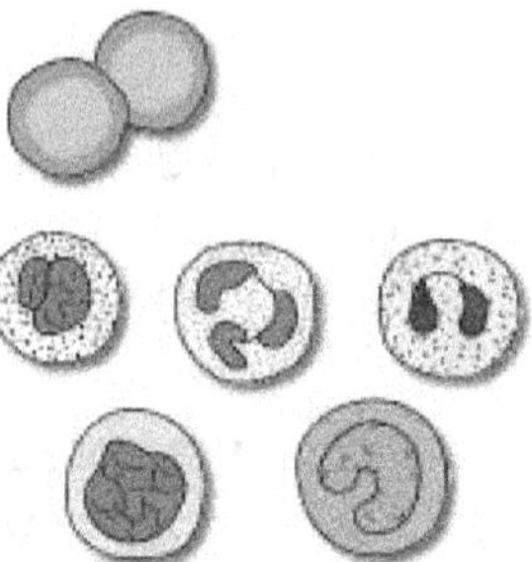

In this example the categorical variable of grade level is the explanatory variable and the dessert preference is the response variable. Notice that the two-way table shows all 9 combinations of the explanatory and the response variables. Each data entry is in its own **cell** of the table.

You may have noticed that the elementary school students have a much higher count (larger ***n***) than the other grade levels. This fact adds difficulty to comparing the different counts.

What if we changed the data into percentages? First we will have to find the totals of each row.

two-way table (dessert preferences)

(grade level)	cupcake	ice cream	candy bar	TOTAL
elementary	336	252	153	741
middle school	77	91	68	236
high school	33	54	162	249
				1226

Now we change the cell count into a **percentage of the grade level** totals.

two-way table (dessert preferences)

(grade level)	cupcake	ice cream	candy bar	TOTAL
elementary	336/741	252/741	153/741	741
middle school	77/236	91/236	68/236	236
high school	33/249	54/249	162/249	249
				1226

(percentages by GRADE LEVEL)

Which goes to …

two-way table (dessert preferences)

(grade level)	cupcake	ice cream	candy bar	TOTAL
elementary	45.3%	34%	20.7%	100%
middle school	32.6%	38.6%	28.8%	100%
high school	13.3%	21.7%	65%	100%

(percentages by GRADE LEVEL)

This table is now easier to see which desserts are favored by grade level. Elementary students seem to prefer cupcakes while high school students prefer candy bars. The middle school students seem to be evenly spread out.

We could have found the percentages by dessert preferences. Here are those calculations.

two-way table (dessert preferences)

(grade level)	cupcake	ice cream	candy bar	
elementary	336	252	153	
middle school	77	91	68	
high school	33	54	162	
TOTAL	446	397	383	1226

(percentages by dessert preferences)

(dessert preferences)

(grade level)	cupcake	ice cream	candy bar	
elementary	$\frac{336}{446}$	$\frac{252}{397}$	$\frac{153}{383}$	
middle school	$\frac{77}{446}$	$\frac{91}{397}$	$\frac{68}{383}$	
high school	$\frac{33}{446}$	$\frac{54}{397}$	$\frac{162}{383}$	
TOTAL				

(percentages by dessert preferences)

(grade level)	cupcake	ice cream	candy bar	
elementary	75.3%	63.5%	39.9%	
middle school	17.3%	22.9%	17.8%	
high school	7.4%	13.6%	42.3%	
TOTAL	100%	100%	100%	

(percentages by dessert preferences)

Again it is easier to see how the desserts are favored by grade level. The results relate the same information as the previous percentages by grade level table (but with different values!). Cupcakes are heavily favored by the elementary students and candy bars are the favorite of high school students. The middle school students still seem to be evenly spread out.

Notice that when we found the percentages by dessert that the ice cream is mostly preferred by the elementary students.

*Please note that we could have reversed the order of the nominal (categorical) variables and put the dessert preferences as the row variables and grade level as the column variables. The results would be the same.

So, how do we determine if there is a significant difference between the grade levels and their dessert preferences? We could perform a z test for each cell, but that only leads to individual cell probabilities instead of an overall significance for the three grade levels.

The main problem with multiple comparisons is that the calculated p-values would belong to each test separately and not to the collective data. The individual p-values would not relate to the **overall** test. Plus, the chances of making a Type I error (rejecting the H_0 when you shouldn't have) increases as the number of comparisons increase. Always remember that we should get significant results around 5% of the time due to chance alone.

Expected cell counts in two-way tables

The null hypothesis always states the same thing: there is no relationship between the categorical variables that label the rows and columns of a two-way table. Typical!

To test the H_0 of a two-way table we need to compare the ***observed*** counts versus the ***expected*** counts. Expected counts are the outcomes that we would expect to occur, including any natural-occurring variance, **if the H_0 were true**. Example: if you flipped a nickel 10,000 times then you would **expect** to get approximately, but not exactly, 5,000 heads and 5,000 tails – if the coin had not been tampered with.

Observed counts (also known as **observations** and/or **measurements**) are actual recorded data in a study. If the observed counts are too far away from the expected counts, then this is strong evidence against the null.

chi-square

Expected counts
Observed counts

Expected cell counts are represented with an **E**.

Observed cell counts are represented with an **O**.

How can we tell if the two counts are **significantly** different? Excellent question. Mgz is proud of you!

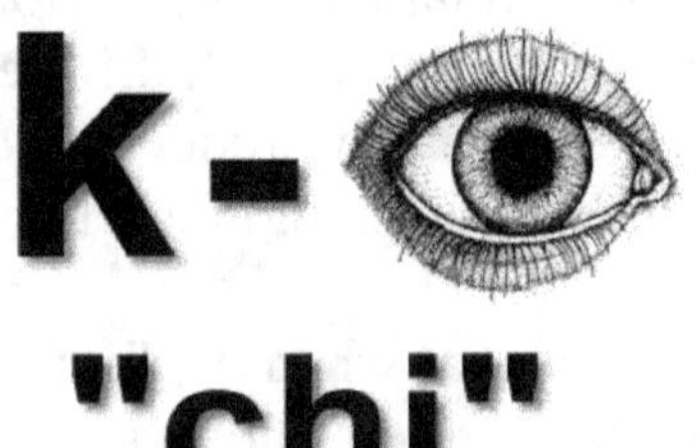

To check for significant differences between two or more categorical variables, we will use a new test for significance: the ***chi-square test***.

Like all tests for statistical significance, we need to crunch the numbers and come up with a **test statistic**, which we then will look up its corresponding p-value and then decide to reject, or not, the null hypothesis. The test statistic generated by a significance test for a two-way table is represented by the Greek letter **chi** (pronounced **k-eye**) which looks like the identical twin to the capital English letter **X**, but with flair.

The first step to finding the chi-square test statistic is to figure out what the **expected counts are for each cell** in the two-way table. This step is not difficult at all, just a lot of multiplying and dividing.

Here is the formula:

Expected counts

If Ho were true, then the *expected* cell count of a two-way table is...

$$\textbf{expected cell count} = \frac{\textbf{row total x column total}}{\textbf{table total}}$$

Example 12.1: Find the expected cell counts (E) from the observed counts (O) in the table below.

Here is the original data with row and column totals:

two-way table (dessert preferences)

(grade level)	cupcake	ice cream	candy bar	TOTAL
elementary	336	252	153	741
middle school	77	91	68	236
high school	33	54	162	249
TOTAL	446	397	383	1226

Here is an example on how to find an expected cell count:

The expected count for this cell is...

	cupcake	ice cream	candy bar	TOTAL
elementary	E	252	153	741
middle school	77	91	68	236
high school	33	54	162	249
TOTAL	446	397	383	1226

$$\frac{446 \times 741}{1226} = 269.56$$

Now we will repeat the same process for each cell.

The Es for all cells.

	cupcake	ice cream	candy bar	TOTAL
elementary	269.56	239.95	231.49	741
middle school	85.85	76.42	73.73	236
high school	90.58	80.63	77.79	249
TOTAL	446	397	383	1226

$$\text{expected cell count} = \frac{\text{row total x column total}}{\text{table total}}$$

Below are both the **Observed** cell counts and the **Expected** cell counts.

Observed cell counts

(grade level)	cupcake	ice cream	candy bar	TOTAL
elementary	336	252	153	741
middle school	77	91	68	236
high school	33	54	162	249
TOTAL	446	397	383	1226

Expected cell counts

	cupcake	ice cream	candy bar	TOTAL
elementary	269.56	239.95	231.49	741
middle school	85.85	76.42	73.73	236
high school	90.58	80.63	77.79	249
TOTAL	446	397	383	1226

Notice the largest differences between the O table and the E table.

Examples: elementary @ cupcake = O – E = 336 – 269.56 = 66.44
elementary @ candy bar = O – E = 153 – 231.49 = -78.49
high school @ cupcake = O – E = 33 – 90.58 = -57.58
high school @ candy bar = O – E = 162 – 77.79 = 84.21

Also notice that half of the differences are positive and the other half are negative, so that is why we **square** them for the chi-square test.

Chi-square test statistic

Now that we know how to find the expected cell counts we can go ahead and perform the chi-square test for significance.

Here is the chi-square test statistic formula:

Chi-square test statistic

$$X^2 = \sum \frac{(\text{observed count} - \text{expected count})^2}{\text{expected count}}$$

$$X^2 = \sum \frac{(O - E)^2}{E}$$

(The $\sum$ means to add up all of the cell differences.)

The X^2 test statistic acts like a standard deviation, or standard error, by telling us how far away the observed counts are from the expected counts.

Chi-square test instructions

1) Calculate the expected counts for each cell of the two-way table.

2) Check the expected counts to make sure that each cell value is greater than 1, and that 80% of the cells are at least 5.

3) Calculate the chi-square test statistic.

$$X^2 = \sum \frac{(O - E)^2}{E}$$

4) Find the degrees of freedom.

two-way table

df = (# of rows - 1)(# of columns - 1)

only one categorical variable

df = (# of variables - 1)

5) Find the corresponding p-value for the calculated X^2 test statistic on the chi-square distribution table.

6) Decide whether to reject the null or not.

There are a few warnings about using a chi-square test. If any of the row or column totals is a zero, then you are going to have a serious problem because when multiplying anything by zero the product is going to be… zero. And of course we cannot divide by zero, not yet anyway but mgz is working on it.

Like all valid statistics analyses, the data needs to be the result of randomly choosing subjects (SRS) – no bias! As always, the larger the sample size the more accurate the statistics. And lastly there is a **general rule that states that every expected cell count should be at least 1 with at least 80% of the cells having a minimum value of 5.**

chi-square

Rules

For a **VALID** chi-square test, the following must be true:

Observed cell counts are from a SRS.

All **expected** cell counts have a value of 1 or more.

At least 80% of **expected** cell counts have a value of at least 5.

Minimum number of degrees of freedom is 1.

Like the t test, the chi-square test uses **degrees of freedom**. The number of degrees of freedom for a two-way table is one less than the total number of rows multiplied by one less than the total number of columns.

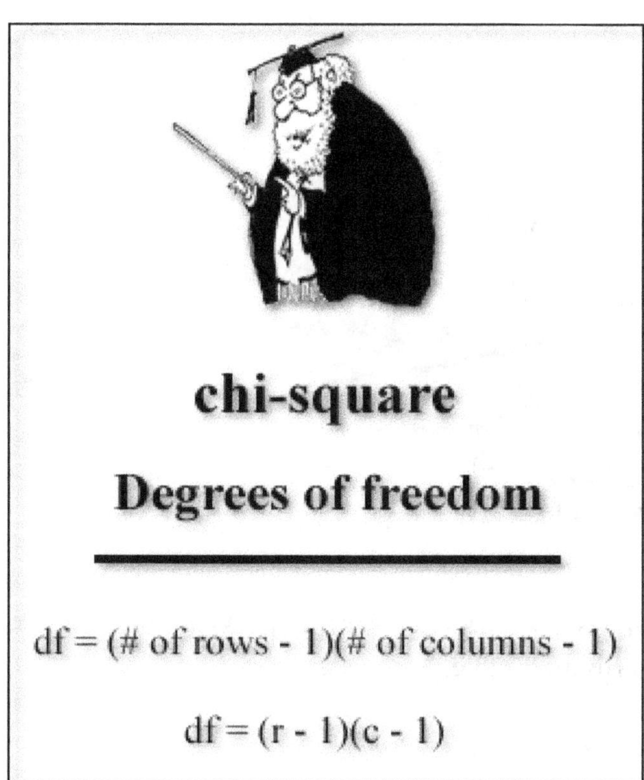

*Important: if there is only one row (or column) then the numbers of degrees of freedom is simply one less than the total number of categorical variables in the row (or column). There are never zero degrees of freedom.

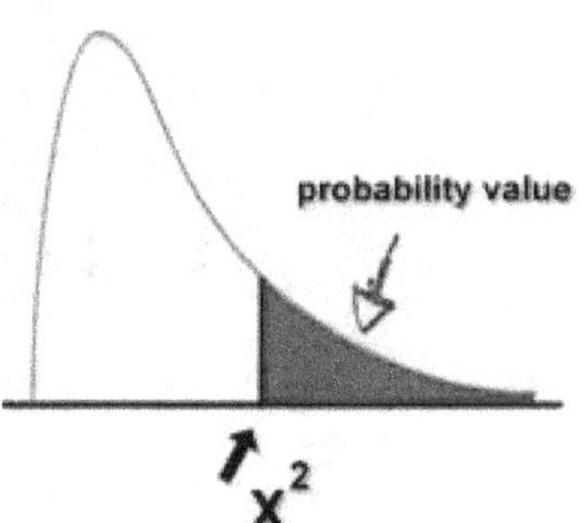

The distributions of the chi-square statistics are always positive (because it is squared!) and is heavily skewed to the right (or positively skewed).

(*The p-value from the chi-square distribution graph is the area to the right of the X^2 statistic.)

Here is the distribution of the chi-square probabilities:

Chi-square distribution

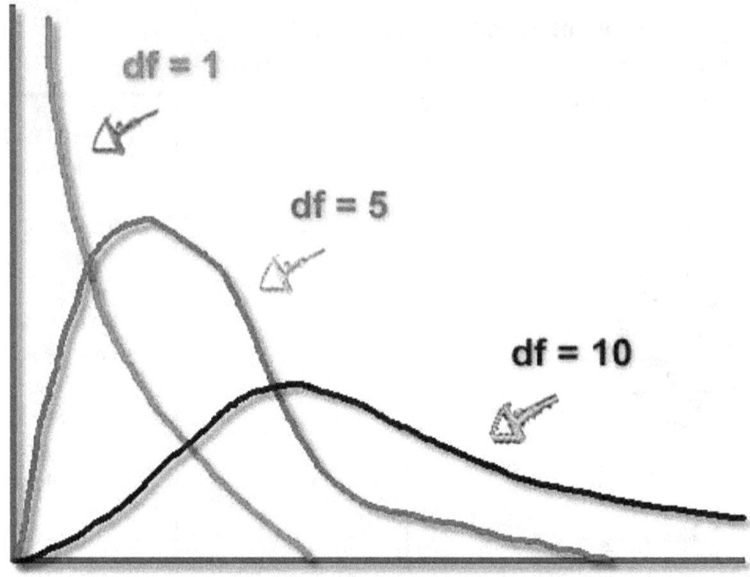

(Notice that the more degrees of freedom, the more normal the distribution.)

After we have calculated a X^2 test statistic, we need to use a different table to find the p-value. We call it the X^2 distribution table. Makes sense to mgz.

Important!

Large test statistics generate small p-values!

The chi-square distribution table is on the next page.

Chi-square (X^2) distribution critical values

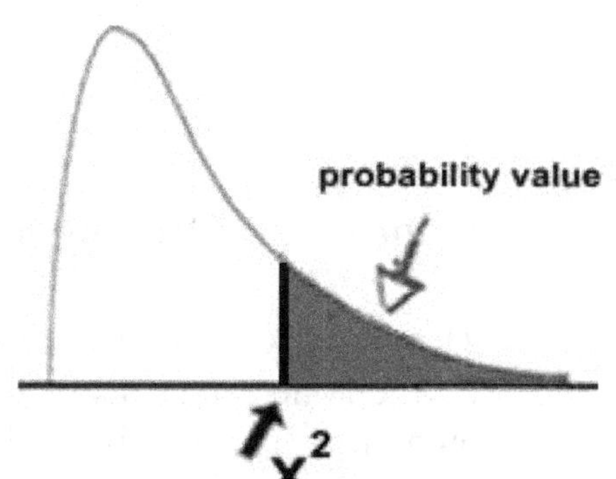

probability values

degrees of freedom

degrees of freedom	0.25	0.20	0.15	0.10	0.05	0.025	0.020	0.010	0.005	0.0025	0.001	0.0005
1	1.32	1.64	2.07	2.71	3.84	5.02	5.41	6.63	7.88	9.14	10.83	12.12
2	2.77	3.22	3.79	4.61	5.99	7.38	7.82	9.21	10.60	11.98	13.82	15.20
3	4.11	4.64	5.32	6.25	7.81	9.35	9.84	11.34	12.84	14.32	16.27	17.73
4	5.39	5.99	6.74	7.78	9.49	11.14	11.67	13.28	14.86	16.42	18.47	20.00
5	6.63	7.29	8.12	9.24	11.07	12.83	13.39	15.09	16.75	18.39	20.52	22.11
6	7.84	8.56	9.45	10.64	12.59	14.45	15.03	16.81	18.55	20.25	22.46	24.10
7	9.04	9.80	10.75	12.02	14.07	16.01	16.62	18.48	20.28	22.04	24.32	26.02
8	10.22	11.03	12.03	13.36	15.51	17.53	18.17	20.09	21.95	23.77	26.12	27.87
9	11.39	12.24	13.29	14.68	16.92	19.02	19.68	21.67	23.59	25.46	27.88	29.67
10	12.55	13.44	14.53	15.99	18.31	20.48	21.16	23.21	25.19	27.11	29.59	31.42
11	13.70	14.63	15.77	17.28	19.68	21.92	22.62	24.72	26.76	28.73	31.26	33.14
12	14.85	15.81	16.99	18.55	21.03	23.34	24.05	26.22	28.30	30.32	32.91	34.82
13	15.98	16.98	18.20	19.81	22.36	24.74	25.47	27.69	29.82	31.88	34.53	36.48
14	17.12	18.15	19.41	21.06	23.68	26.12	26.87	29.14	31.32	33.43	36.12	38.11
15	18.25	19.31	20.60	22.31	25.00	27.49	28.26	30.58	32.80	34.95	37.70	39.72
16	19.37	20.47	21.79	23.54	26.30	28.85	29.63	32.00	34.27	36.46	39.25	41.31
17	20.49	21.61	22.98	24.77	27.59	30.19	31.00	33.41	35.72	37.95	40.79	42.88
18	21.60	22.76	24.16	25.99	28.87	31.53	32.35	34.81	37.16	39.42	42.31	44.43
19	22.72	23.90	25.33	27.20	30.14	32.85	33.69	36.19	38.58	40.88	43.82	45.97
20	23.83	25.04	26.50	28.41	31.41	34.17	35.02	37.57	40.00	42.34	45.31	47.50
21	24.93	26.17	27.66	29.62	32.67	35.48	36.34	38.93	41.40	43.78	46.80	49.01
22	26.04	27.30	28.82	30.81	33.92	36.78	37.66	40.29	42.80	45.20	48.27	50.51
23	27.14	28.43	29.98	32.01	35.17	38.08	38.97	41.64	44.18	46.62	49.73	52.00
24	28.24	29.55	31.13	33.20	36.42	39.36	40.27	42.98	45.56	48.03	51.18	53.48
25	29.34	30.68	32.28	34.38	37.65	40.65	41.57	44.31	46.93	49.44	52.62	54.95
26	30.43	31.79	33.43	35.56	38.89	41.92	42.86	45.64	48.29	50.83	54.05	56.41
27	31.53	32.91	34.57	36.74	40.11	43.19	44.14	46.96	49.64	52.22	55.48	57.86
28	32.62	34.03	35.71	37.92	41.34	44.46	45.42	48.28	50.99	53.59	56.89	59.30
29	33.71	35.14	36.85	39.09	42.56	45.72	46.69	49.59	52.34	54.97	58.30	60.73
30	34.80	36.25	37.99	40.26	43.77	46.98	47.96	50.89	53.67	56.33	59.70	62.16
40	45.62	47.27	49.24	51.81	55.76	59.34	60.44	63.69	66.77	69.70	73.40	76.09
50	56.33	58.16	60.35	63.17	67.50	71.42	72.61	76.15	79.49	82.66	86.66	89.56
60	66.98	68.97	71.34	74.40	79.08	83.30	84.58	88.38	91.95	95.34	99.61	102.69
80	88.13	90.41	93.11	96.58	101.88	106.63	108.07	112.33	116.32	120.10	124.84	128.26
100	109.14	111.67	114.66	118.50	124.34	129.56	131.14	135.81	140.17	144.29	149.45	153.17

Goodness of fit chi-square test

There are two main ways to use a chi-square test. A **"goodness of fit"** test is when one categorical variable is compared to the overall expected outcomes. This test is similar to comparing a sample mean against a stated population mean.

A **"test of independence"** is used to see if one categorical variable is "**related**" to another categorical variable. In other words, is one variable ***dependent*** on another variable? This test is similar to a two-sample t test.

A ***goodness of fit*** test will tell us if a single cell count differs significantly from the other cell counts. In other words, if one individual categorical variable is *significantly different* from the overall probabilities from all of the categorical variables.

Example 12.2: Below is data from the National Center for Health Statistics. Is one month significantly different from the other months for the number of births in the US?

U.S. Births by Month, 1995-2002 (in thousands)	Total Births
January	2,582
February	2,410
March	2,645
April	2,538
May	2,674
June	2,629
July	2,789
August	2,814
September	2,741
October	2,695
November	2,532
December	2,632
Total1995-2002	31,679

Source: Calculated from National Center for Health Statistics data

Step 1) What is the H_0?

The categorical variables are the months. So, we need to pretend that the probability of a birth in any month is the same.

Jan	Feb	Mar	Apr	May	Jun	Jul	Aug	Sep	Oct	Nov	Dec
p1	p2	p3	p4	p5	p6	p7	p8	p9	p10	p11	p12

Therefore the null hypothesis would be…

$\mathbf{H_0: p_1 = p_2 = p_3 = p_4 = p_5 = p_6 = p_7 = p_8 = p_9 = p_{10} = p_{11} = p_{12} = 1/12}$

What is the H_a? The alternative hypothesis states that at least one of the months is significantly different from the other months.

$\mathbf{H_a}$**: not all** $\mathbf{p_i = 1/12}$

Step 2) Find the chi-square statistic: (This is where the work is!)

Here is the X^2 test statistic formula:

$$X^2 = \sum \frac{(O - E)^2}{E}$$

Find the **expected counts**. Since the null states that all probabilities are the same, then if the null were true, all of the month births would be identical. We divide the total number of births by 12 and that makes the expected counts by each month: 31,679/ 12 $\approx$ 2,640 (thousands).

(in thousands)	Observed counts (O)	Expected counts (E)
January	2,582	2,640
February	2,410	2,640
March	2,645	2,640
April	2,538	2,640
May	2,674	2,640
June	2,629	2,640
July	2,789	2,640
August	2,814	2,640
September	2,741	2,640
October	2,695	2,640
November	2,532	2,640
December	2,632	2,640

Now use the formula to find the chi-square test statistic.

(in thousands)	Observed counts (O)	Expected counts (E)	O - E	$(O - E)^2$	$(O - E)^2/E$	
January	2,582	2,640	-58	3,364	1.27	
February	2,409	2,640	-231	53,361	20.11	
March	2,645	2,640	5	25	0.01	
April	2,537	2,640	-103	10,609	3.95	
May	2,673	2,640	33	1,089	0.44	
June	2,629	2,640	-11	121	0.04	
July	2,788	2,640	148	21,904	8.38	
August	2,813	2,640	173	29,929	11.42	
September	2,740	2,640	100	10,000	3.85	
October	2,694	2,640	54	2,916	1.13	
November	2,532	2,640	-108	11,664	4.40	
December	2,631	2,640	-9	81	0.03	
					55.03	← X^2

The calculated X^2 test statistic is **55.03.**

Step 3) Find the number of degrees of freedom (only 1 row with no columns).

df = (# of variables in row – 1) = 12 – 1 = 11

Step 4) Find the corresponding p-value from the X^2 distribution table for a chi-square test statistic of **55.03** with **11** degrees of freedom.

probability values

		0.25	0.20	0.15	0.10	0.05	0.025	0.020	0.010	0.005	0.0025	0.001	0.0005	p < .0005
	1	1.32	1.64	2.07	2.71	3.84	5.02	5.41	6.63	7.88	9.14	10.83	12.12	
	2	2.77	3.22	3.79	4.61	5.99	7.38	7.82	9.21	10.60	11.98	13.82	15.20	
	3	4.11	4.64	5.32	6.25	7.81	9.35	9.84	11.34	12.84	14.32	16.27	17.73	
	4	5.39	5.99	6.74	7.78	9.49	11.14	11.67	13.28	14.86	16.42	18.47	20.00	
	5	6.63	7.29	8.12	9.24	11.07	12.83	13.39	15.09	16.75	18.39	20.52	22.11	
	6	7.84	8.56	9.45	10.64	12.59	14.45	15.03	16.81	18.55	20.25	22.46	24.10	X^2
	7	9.04	9.80	10.75	12.02	14.07	16.01	16.62	18.48	20.28	22.04	24.32	26.02	
	8	10.22	11.03	12.03	13.36	15.51	17.53	18.17	20.09	21.95	23.77	26.12	27.87	
	9	11.39	12.24	13.29	14.68	16.92	19.02	19.68	21.67	23.59	25.46	27.88	29.67	
	10	12.55	13.44	14.53	15.99	18.31	20.48	21.16	23.21	25.19	27.11	29.59	31.42	
11 dfs	11	13.70	14.63	15.77	17.28	19.68	21.92	22.62	24.72	26.76	28.73	31.26	33.14	55.03

Looking up p-values on a chi-square distribution table is similar to looking up a t test statistic probability value. Find the correct degrees of freedom row and then find the two critical values that "bracket" the calculated chi-square test statistic, then the corresponding p-value is between the two probability values at the top of the bracketing columns.

Notice that our X^2 test statistic of 55.03 is greater than the last critical value (33.14) in the 11th df row. That simply means that the corresponding p-value to our calculated chi-square test statistic of 55.03 is less than the probability value of the last column (.0005).

So our p-value for a X^2 test statistic of 55.03 with 11 df is **p < .0005**. (*Most software programs simply list very small p-values like ours as p < .001.)

Step 5) Decide to reject, or not reject, the null hypothesis.

Since our calculate p-value is far less than the critical alpha of .05, **we reject the null**. (Small p-values are bad news for the null.)

Step 6) Interpret:

There is a significant difference in the number of US births by month.

Test for independence chi-square test

The chi-square test can also be used to ***test for independence*** between categorical variables that label the rows and columns of a two-way table.

A **"test of independence"** is used to see if one categorical variable is "**related**" to another categorical variable. In other words, is one variable ***dependent*** on another variable? This test is similar to a two-sample t test.

Example 12.3: An election poll reported the following breakdown of political parties by gender:

test for independence

(party)	(gender) male	female
Democratic	**4125**	**3866**
Republican	**5836**	**5102**

So, according to this two-way table, is there any relationship between a person's gender and their political party affiliation? We can also state this question as "*are gender and political party affiliation independent*?" In other words, are the column categorical variables (gender) independent of the row categorical variables (political party)?

Step 1) What are the null and alternate hypotheses?

The null states that there is no relationship which translates into…

H_0: gender is **independent** of political party (no relationship)

H_a: gender is **dependent** of political party (there is a relationship)

Step 2) Find the expected cell counts.

test for independence **Observed** cell counts

(gender)

(party)	male	female	TOTAL
Democratic	4125	3866	7991
Republican	5836	5102	10938
TOTAL	9961	8968	18929

Goes to…

test for independence **Expected** cell counts

(gender)

(party)	male	female
Democratic	$\frac{7991 \times 9961}{18929}$	$\frac{7991 \times 8968}{18929}$
Republican	$\frac{9961 \times 10938}{18929}$	$\frac{8968 \times 10938}{18929}$

Goes to…

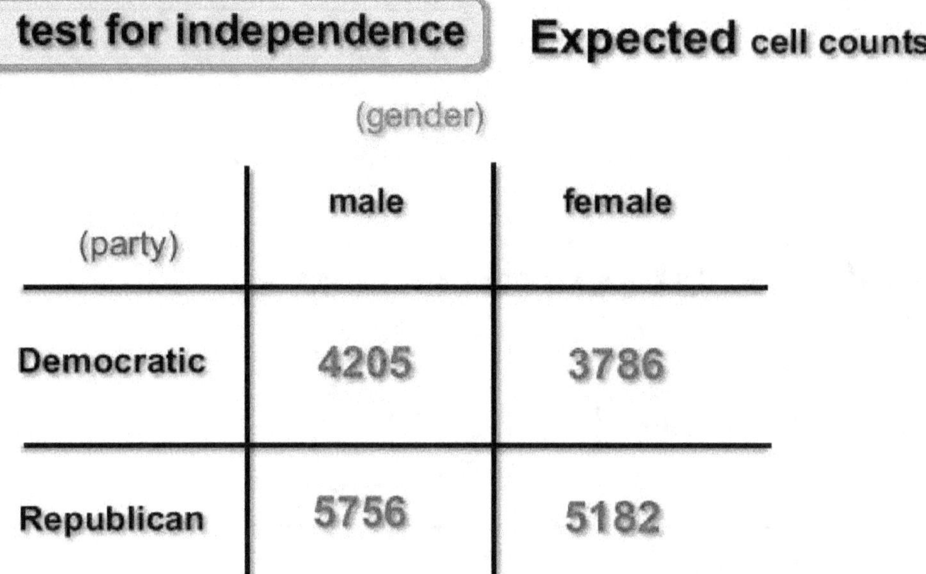

test for independence Expected cell counts

(party)	male (gender)	female
Democratic	4205	3786
Republican	5756	5182

Step 3) Break out the X^2 test statistic formula and calculate each cell difference.

$$X^2 = \sum \frac{(O - E)^2}{E}$$

		O	E	O - E	$(O - E)^2$	$(O - E)^2/E$	
cell 1,1	(Dem males)	4125	4205	-80.10	6416.12	1.53	
cell 1,2	(Rep males)	5836	5756	80.10	6416.12	1.11	
cell 2,1	(Dem females)	3866	3786	80.10	6416.12	1.69	
cell 2,2	(Rep females)	5102	5182	-80.10	6416.12	1.24	
						5.57	← X^2

Step 4) Find the degrees of freedom.

df = (r – 1)(c – 1) = (2 – 1)(2 – 1) = 1.

Step 5) Look up the probability value from the calculated X^2 test statistic in the chi-square distribution table.

So, the p-value for a X^2 test statistic of 5.57 with 1 degree of freedom is …

Between .02 and .01… (.01 < p-value < .02).

Step 6) Interpret:

Since our p-value is less than .05, we reject the null hypothesis. From this data, we can safely state that there is a significant relationship between gender and political party affiliation.

Chapter 12 Review

A ***two-way table*** is a representation of data from two or more categorical variables with **rows** and **columns.** A ***cell*** is a single "box" in a two-way table that holds data of where the two categorical variables intersect.

Observed counts (also known as **observations** and/or **measurements**) are actual recorded data in a two-way table. ***Expected*** counts are calculated values in a two-way table.

Expected counts

If Ho were true, then the *expected* cell count of a two-way table is...

$$\text{expected cell count} = \frac{\text{row total x column total}}{\text{table total}}$$

The ***chi-square (X^2) test statistic*** acts like a standard deviation, or standard error, by telling us how far away the observed counts are from the expected counts.

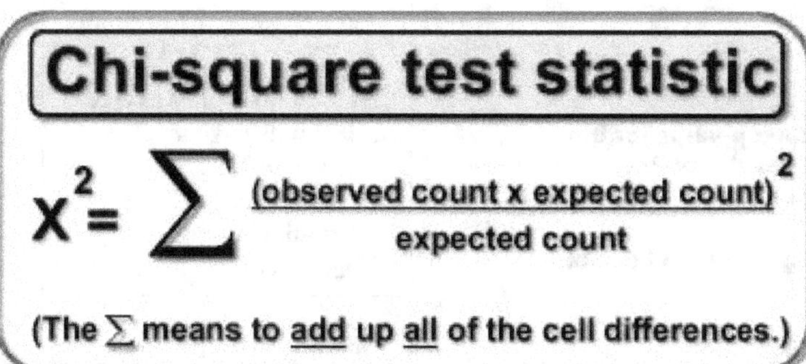

Degrees of freedom are used in a chi-square test.

chi-square degrees of freedom.

two-way table
df = (# of rows - 1)(# of columns - 1)

only one categorical variable
df = (# of variables - 1)

There are two main ways to use a chi-square test. A **"*goodness of fit*"** test is when one categorical variable is compared to the overall expected outcomes. A **"*test of independence*"** is used to see if one categorical variable is **dependent** on another categorical variable.

Instructions for a chi-square test:

Chi-square test instructions

1) Calculate the expected counts for each cell of the two-way table.

2) Check the expected counts to make sure that each cell value is greater than 1, and that 80% of the cells are at least 5.

3) Calculate the chi-square test statistic.

$$X^2 = \sum \frac{(O - E)^2}{E}$$

4) Find the degrees of freedom.

two-way table
df = (# of rows - 1)(# of columns - 1)

only one categorical variable
df = (# of variables - 1)

5) Find the corresponding p-value for the calculated X^2 test statistic on the chi-square distribution table.

6) Decide whether to reject the null or not.

chi-square Rules

For a **VALID** chi-square test, the following must be true:

Observed cell counts are from a SRS.

All **expected** cell counts have a value of 1 or more.

At least 80% of **expected** cell counts have a value of at least 5.

Minimum number of degrees of freedom is 1.

Guided homework Chapter 12: Chi-square tests

1. The State of California rates their schools using an Annual Performance Index rating. Five high schools from the same school district reported their 2010 API school ratings:

School	A	B	C	D	E
API score	733	687	729	731	749

Use a chi-square test to determine if there is a significant difference between the schools' API scores.

2. A local high school gave a final exam in intermediate algebra to all of their 11th graders. The test grades are broken down into gender categories.

11th grade		males	females
	A	14	12
	B	32	38
	C	67	76
	D	86	93
	F	124	107

a) How many males took the test?

b) What percent of the female students earned a C on the test?

c) What percent of all students earned less than a C on the test?

d) What is the expected cell count for male students who earned an A on the test?

e) What is the chi-square statistic?

f) What are the degrees of freedom?

g) Is there a significant difference between the grading of male and female students?

3. In a survey 200 people were asked to rate whether they felt: more relax, the same, or more stressed after listening to 5 minutes for each of two pieces of music (jazz and classical). The results were as follows:

		State of stress			
		More relaxed	*The same*	*More stress*	
Music genre	*Jazz*	**27**	**22**	**42**	
	Classical	**48**	**40**	**21**	

Using this data, find out if there is a relationship between music genre and state of stress.

Chapter 12 Quiz

1. Julie is in charge of the music that the passengers of a cruise liner listen to in their shipboard quarters. She took a survey of listening preferences and recorded the data below. Use a goodness-of-fit chi-square test to verify if there is a significant difference between the genres.

Genre of radio	classic rock and roll	jazz	religious	classics	talk radio
#	41	22	20	28	36

2. A new movie company is trying to decide where to invest their money. They did a quick random poll at a local movie theater to see if there was a significant difference between the genders when it came to movie genres. Here is the data. Use a test-for-independence chi-square test to see if there is a significant relationship between genre of movie and gender.

observed		**males**	**females**
	Horror	36	25
	Sci-fi	48	47
	Crime	34	29
	Comedy	45	44
	Drama	17	41

3. Is there a difference between the political parties when it comes to taxation? The two-way table recorded a survey of randomly chosen voters who self-identified as either democrat or republican. Is there? Prove your answer.

	More taxes	**No new taxes**	**Less taxes**
Republican	**550**	**376**	**231**
Democrat	**495**	**352**	**244**

Final exam: Chapters 6 - 12

Final

Exam

1. A recent study showed that the probability of randomly chosen homeowners has the probability of 0.80 that the value of their house had gone down in value over the previous year. Let's pretend that you were going to call 1,000 of these homeowners at random to find out how many of them lost value in their homes.

a) How many of them (out of the thousand) would you expect to report a loss in value? (Hint: what is the mean?)

b) What would the standard deviation be?

c) What is the probability that you will get exactly 23 homeowners that lost value?

2. If you built a 90% confidence interval, then the critical value alpha (α) would be what?

3. The average height for males between the ages of 25 and 39 are normally distributed with a mean of 69 inches and a standard deviation of 3.6 inches. According to the Central Limit Theorem, if you took sample of 100 men between the ages of 25 and 39 at random, what would the standard deviation be?

4. Which of the following is the corresponding level of confidence C for a critical (z*) value of 1.960?

a. 99%
b. 62.5%
c. 95%
d. 90%
e. 100%

5. True or false? - The test statistic is not the same as the probability value, but it does have a corresponding probability value from the normal distribution curve.

Final

Exam

6. Ornithologists claim that the Ruby-throated hummingbird beats its wings 60 times per second with a standard deviation of 21.5 beats. Harvey Birdman disagrees because he filmed 50 of the lightning birds in flight on a special camera and he came up with an average of 66 wing flaps per second.

a) What is the null hypothesis?

b) What are the two possible alternative hypotheses?

c) What is the test statistic?

d) What is the p-value if this is a one-tailed test?

e) What is the p-value if this is a two-tailed test?

f) Assuming we will use a two-tail test, is the data significant at $\alpha = .05$?

g) Assuming we will use a two-tail test, is the data significant at $\alpha = .01$?

7. The Central Limit Theorem states that when you take a sample from a population, go ahead and use the sample mean instead of the population mean in the normal distribution formulas, and then divide the population _____ _____ by the square root of the sample size.

a. stated mean
b. average median
c. binomial probability
d. general probability
e. standard deviation

8. True or false? - If the stated, or claimed, population mean is indeed true, then the sampling means will fall between the confidence interval values 90% of the time, 95% of the time, or whatever percent of the time that corresponds to the level of confidence C.

9. A farmer harvests his banana crop and weighs them in grams. Their weights fit a normal distribution of N(206, 18). What is the probability that:

a) One banana selected at random will weigh over 250 grams?

b) Four bananas will have an average weight of less than 188 grams?

c) One hundred bananas will have a mean weight between 205 and 209 grams?

d) One thousand bananas will have a mean weight of more than 207 grams?

e) Ten thousand bananas will have a mean weight of more than 206 grams?

10. A kangaroo-ologist weighed 50 four-month old joeys and got a sample mean of 18.6 lbs with a standard deviation of 1.86 lbs. He wants to estimate the true mean weight of four-month old joeys with a margin of error of 0.1 lbs with a 90% confidence interval. How many joeys should he weigh?

11. The statistical science that describes a set of measurements, or observations, is called ________ statistics.

a. inferential
b. descriptive
c. parametric
d. responsive
e. quantitative

12. A new drug claims to make soldiers shoot straighter. A group of 8 soldiers are given the drug and another group of 8 were not given the drug. Both groups were taken to the firing range and the table below lists the number of bull's-eyes each soldier scored.

Red	Blue
12	12
10	13
11	14
14	14
10	12
11	13
13	14
12	15

Final

Exam

a) What is the null hypothesis?

b) What is the alternative hypothesis?

c) What is the test statistic?

d) What is the p-value?

e) Is the data significant at $\alpha = .05$?

f) Is the data significant at $\alpha = .01$?

13. What are you confident about with a 95% confidence interval?

a. A margin of error that equals 95%
b. That if the experiment was repeated many times, then 95% of the time the sample means would fall between these two values
c. That you are 95% sure that the mean will fall between the two values
d. 95% of the measurements will be less than the mean
e. 95% of the data will fall between the two values

14. The average weight of adult male rhinoceri (the plural of rhinoceros can be rhinoceros, rhinoceri, rhinoceroses, or rhinoceroi) is claimed to be 4500 lbs with a standard deviation of 116 lbs. A zoologist is able to weigh 35 adult male rhinoceri from a crash of rhinos. A crash is a group of rhinos...weird eh? Who comes up with these group names anyway? The zoologist gets a sample mean of 4527 lbs and writes back to his zoological colleagues that the average weight of the rhinos should be officially changed to a greater weight. (Assume this is a one-tailed test.)

a) What is the null hypothesis?

b) What is the alternative hypothesis?

c) What is the test statistic?

d) What is the alternative hypothesis?

e) What is the p-value?

f) Is the data significant at the level of .10?

g) Is the data significant at the level of .05?

Final
Exam

15. True or false? - Measures of central tendency from populations are represented with letters from the Greek alphabet.

16. True or false? - Binomial probability formulas must have an infinite number of events or trials.

17. Which of the following is not a binomial setting?

a. The probability of picking 5 numbers from a field of 39 numbers
b. The probability of getting 10 heads out of 10 coin tosses
c. The probability of randomly picking three blue socks from a drawer that has three red socks, 8 black socks, 16 white socks and 12 blue socks
d. The probability of getting an odd number on a roulette wheel
e. All of these are binomial settings

18. True or false? - The random sample space S of a binomial setting is the sample size squared.

Final

Exam

19. True or false? - When building a confidence interval, it is extremely important that the sample means fit a normal distribution curve.

20. True or false? - Testing a statement for accuracy with statistics is called testing the alternative hypothesis.

21. When would you use a one-sample t test instead of a z test?

a. when the sample size is 30 or less and the population standard deviation is unknown
b. when the sample size is greater than 30 and the population standard deviation is known
c. when the sample size is over 30 and the population mean is unknown
d. when the sample size is less than 30 and the probability is unknown
e. when the population mean is unknown

22. True or false? alpha (α) + confidence interval = 100%

23. Retired San Diego Padre Tony Gwinn Sr. has a lifetime batting average of .338. WOWZERS!!! That means that he got a hit close to once out of every three at bats. DOUBLE WOWZERS!!! That is impressive indeed. So, what were the chances that he would get 5 hits out of 5 at bats?

24. Not-so-jolly Roger the pirate is testing the hypothesis that the average weight of buried treasure is 865 lbs, as was stated in the last issue of Pirates Today magazine. Roger steals 15 buried treasures from other pirates (who he believes are scurvy dogs anyway) and gets a sample mean of 941 lbs with a standard deviation of 175 lbs. (Assume this is a two-tailed test.)

a) Is the data significant at the .01 level?

25. The Law of Large Numbers basically states that:

a. the larger the percentage, the better the odds
b. the probability is larger
c. the larger the sample size the more accurate the statistics will be
d. the mean will be larger
e. some numbers grow very large and need to be put on a diet

26. Construct a 90% confidence interval for a sample of data that has a sample mean of 116 grams with a known population standard deviation (σ) of 23 grams and a sample size of 400.

27. Build a 98% confidence interval from a sample of 50 with a sample mean of 165.8 and a population standard deviation of 27.6.

28. A confidence interval is the ______ _______ plus/minus the margin of error.

a. variance
b. standard deviation
c. level of confidence
d. critical value
e. estimate

29. California has a Fantasy Five lottery game where the gambler picks five numbers from 1 - 39. What are the odds that a single ticket will have all five winning numbers?

Final

Exam

30. Construct a 95% confidence interval from a sample size of 69 with a sample mean of 2.00 and a population standard deviation of 0.033.

31. True or false? – Normally, the number of degrees of freedom is equal to one less that the sample size.

32. Which of the following definitions fits best for "random variable X" in a binomial probability problem?

a. The probability of a successful outcome
b. The count of successful outcomes
c. The sample size multiplied by the probability of a binomial setting
d. The binomial setting
e. The mean of a binomial setting

33. Make a margin of error for a 90% confidence level with a standard deviation of 18.7 and a sample size of 75.

34. What size confidence interval would you construct if you had a critical value alpha (α) of .1?

35. Rudy wants to take advantage of any four topping deal at the pizza parlor. How many different four topping pizzas are there if the pizza restaurant offers 15 different toppings?

36. The National Basketball Association claims that the average shoe length for their players is 13 inches. Al the Shoe Man believes that the average shoe size is larger that 13 inches so he sneaks into the locker room of the Denver Spurs and measures all the shoes he can find. The table below lists his results in inches. (Assume this is a one-tailed test.)

12.5	12
11.9	12.1
13.5	11.7
14.1	11.4
13.2	13.2
14.3	14.1
11.9	12.4

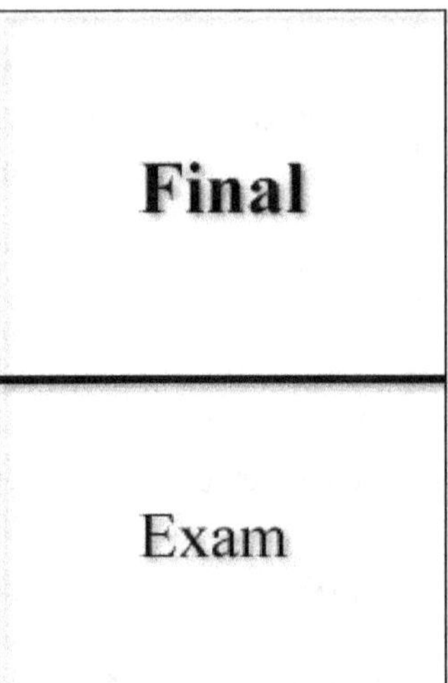

a) What is the null hypothesis?

b) What is the alternative hypothesis?

c) What is the test statistic?

d) What is the p-value?

e) Is the data significant at $\alpha = .05$?

f) Is the data significant at $\alpha = .01$?

37. True or false? - A standard error is a standard deviation from a sample. (Assume that you will divide it by the square root of the sample size, yes?)

38. True or false? - A student tests positive for the Swine Flu when in reality she does not have the flu is an example of a Type II error.

39. True or false? - Assume you are testing a hypothesis that states $\mu = 12.5$ against a sample mean of 12.1 with a standard deviation of 2.39. Then the hypotheses would look like this: H_0: u = 12.5 and H_a: u = 12.1.

40. True or false? - If the critical value alpha (α) is not mentioned, we always use the value of .05.

41. Back in WWII, British bomber pilots had a survival rate of 95% for every mission that they flew. What are the chances that a single pilot would survive long enough to fly 50 missions?

a. Around 32.7%
b. Around 5.6%
c. Around 7.7%
d. Around 24.1%
e. Around 3.4%

Final
Exam

42. True or false? - A Type I error would be not rejecting the null hypothesis when it was actually false.

43. True or false? - A two-sample t test analyzes the difference between a sample group and a stated, or claimed, population average while a one-sample t test checks for a significant difference between two groups with 30 or less individuals.

44. If there is enough evidence to reject the __________ we call the data significant at the alpha level.

a. probability interval
b. confidence interval
c. binomial setting
d. alternative hypothesis
e. null hypothesis

45. Which of the following is the corresponding critical (z*) value for a 99% level of confidence C?

a. 1.558
b. 1.645
c. 1.960
d. .95
e. 2.576

46. The person who runs a high school cafeteria recorded that on Thursday's lunch that 671 students out of 1200 students chose the taco plate for their lunch. What is the sample proportion of students that had the taco plate?

a. p-hat = 671
b. p-hat = 0.671
c. p-hat = 0.559
d. p-hat = 1.788
e. p-hat = 1200

47. In 2010 a government report stated that 44.5% of all registered voters in Californian were democrats. Maria works for a local politician who asks her to take her own poll to verify, or not, if that proportion is true for the politician's district. She hangs out in front of a major grocery store with low low prices and asks shoppers if they were a registered voter and a democrat. Out of the 234 people who said that there were registered voters, 108 said that they were democrats. From Maria's data, which of the following is correct about her sample proportion analysis?

a. H0: p_0= .445 Ha: $p_0 \neq .44$, p-hat = 0.4616, SE = 0.0325, z = 0.51, p-value = .61, do not reject null
b. H0: p_0= .445 Ha: $p_0 \neq .445$, p-hat = 0.4616, SE = 0.2114, z = 1.83, p-value = .025, reject null
c. H0: p_0= .445 Ha: $p_0 = .462$, p-hat = 0.445, SE = 0.04092, z = -1.54, p-value = .86, do not reject null
d. H0: p_0= .445 Ha: $p_0 \neq .445$, p-hat = 0.445, SE = 0.0947, z = 2.12, p-value = .005, reject null
e. H0: p_0= .445 Ha: $p_0 \neq .445$, p-hat = 0.462, SE = 0.6833, z = -0.54, p-value = .95, do not reject null

48. Ms. Johnson, the Vice Principal at a high school, took a survey of all of the students who participated in a school-sanctioned sport about which class was their favorite. She used a goodness of fit chi-square test. What is the chi-square test statistic?

class	English	Math	History	Biology	PE
observed	24	9	25	24	29

a. -7.15
b. 12.09
c. 0.03
d. 6.32
e. 10.58

Final

Exam

49. A huge cable company is beginning a new marketing campaign to increase the number of subscribers for their pay channels. The data was recorded below. A test for independence chi-square test was done to see if there was a relationship between gender and pay channel preferences. Which of the following is true?

observed		males	females
	HBO	54	35
	Showtime	36	32
	Cinemax	34	26
	Epix	45	51
	Disney	21	37

a. $X^2 = 6.62$, df = 4, p-value > .05, do not reject null
b. $X^2 = 10.09$, df = 4, p-value < .05, reject null
c. $X^2 = 5.48$, df = 4, p-value > .05, do not reject null
d. $X^2 = 12.24$, df = 4, p-value <.01, reject null
e. $X^2 = 9.93$, df = 4, p-value < .05, reject null

50. Does the hand that a person eats with influence the flavor of ice cream that a person prefers? A survey was done at an ice cream parlor.

	chocolate	vanilla	strawberry
right-handed	124	112	63
left-handed	15	19	17

A test for independence chi-square test was done to see if there was a relationship between handedness and ice cream preferences.
Which of the following is true?

a. $X^2 = 3.09$, p-value > .05, do not reject null
b. $X^2 = 9.73$, p-value < .05, reject null
c. $X^2 = 6.48$, p-value < .05, reject null
d. $X^2 = 4.46$, p-value >.05, do not reject null
e. $X^2 = 9.93$, p-value < .05, reject null

Final

Exam

Index

Index
Index

Index
Index

Index

Index

Index

Index

Index

Index

Index

Index

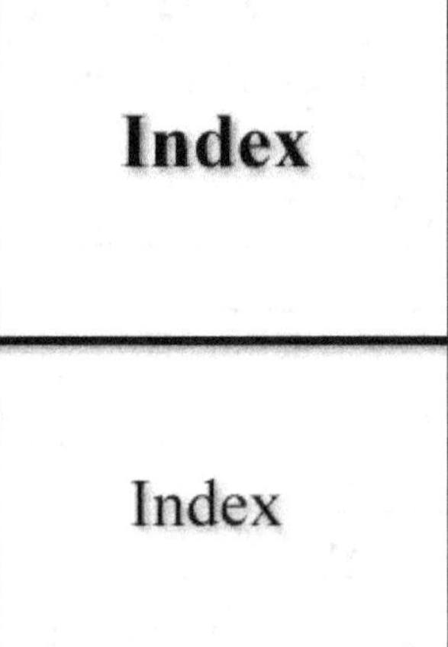
Index
Index

Appendices

Excel formulas

Mean	**=AVERAGE(A1:A10)**

Median	**=MEDIAN(A1:A10)**

Mode	**=MODE(A1:A10)**

First quartile (lower quartile)	
	=QUARTILE(A1:A10,1)

Third quartile (upper quartile)	
	=QUARTILE(A1:A10,3)
Minimum	**=MINIMUM(A1:A10)**
Maximum	**=MAXIMUM(A1:A10,3)**

Variance	**=VAR(A1:A10)**
Standard Deviation	**=STDEV(A1:A10)**
Correlation	**=CORREL(A1:A10, B1:B10)**

To calculate a standardized score (z score)

z score of random variable x with mean μ and standard deviation σ (A and B are arbitrary numbers.)	
P(x < A)	**=STANDARDIZE(x, μ, σ)**

Finding probabilities using a standardized Score (z score)

The z values represent specific percents of probabilities (A and B are arbitrary numbers.)	
z < A	**=NORMSDIST(A)**
z > A	**=1-NORMSDIST(A)**
A < z < B	**=NORMSDIST(B)-NORMSDIST(A)**

Finding the random variable given a percentage first

Value of x with a specified percent % (or proportion) p <u>below</u> it
=NORMINV(p%, μ,σ)
Value of X with a specified percent % (or proportion) p <u>above</u> it
=NORMINV(1-p%, μ,σ)

Finding cumulative probabilities using the random variable x

Probabilities of x with mean μ and standard deviation σ
(A and B are arbitrary numbers.)

P(x < A)	**=NORMDIST(A,μ,σ,TRUE)**
P(x > A)	**=1-NORMDIST(A,μ,σ,TRUE)**
P(A < x < B)	=NORMDIST(B,μ,σ,TRUE) - NORMDIST(A,μ,σ,TRUE)

Finding cumulative probabilities using the sample mean x

Probabilities of x with mean μ and standard deviation σ
(A and B are arbitrary numbers.)

P(x < A)	=NORMDIST(A,μ,σ/√n,TRUE)
P(x > A)	=1-NORMDIST(A,μ,σ√n,TRUE)
P(A < x < B)	=NORMDIST(B,μ,σ√n,TRUE)-NORMDIST(A,μ,σ√n,TRUE)

Binomial probabilities

P(A = k) <u>exactly</u> k successes out of *n* trials with probability *p* of success for each outcome (exact, not cumulative)
=BINOMDIST(A,n,p,false)

P(A ≤ k) <u>at least</u> k successes out of n trials with probability p of success for each outcome (cumulative)
=BINOMDIST(A,n,p,true)

Confidence intervals

Build a CI with confidence level C (Remember that 100% - CI = α.)
=CONFIDENCE(α,σ,n)

t tests

UNKNOWN σ *one sample* with degrees of freedom of n *-1*

Let X = absolute value of the t test statistic

Probability (p-value) **One tail =TDIST(X, df, 1)**

Probability **Two tail =TDIST(X, df, 2)**

t tests

UNKNOWN σ *two sample* with degrees of freedom of n1 *-1* or n_2 *-1* ***(Whichever value is SMALLER!!!!)***

Let X = absolute value of the t test statistic

Probability **One tail =TDIST(X, df, 1)**

Probability **Two tail =TDIST(X, df, 2)**

t test probability (two-sample)
=TTEST(A1:A10, B1:10, # tails, type)

chi-square tests

chi-square distribution p-value **=CHIDIST(X^2, df)**

chi-square test **=CHITEST(Observed array, Expected array)**

Statistical Formulas

Statistical formulas

mean

$$\bar{X} = \frac{\sum x_i}{n}$$

correlation

$$r = \frac{1}{n-1}\Sigma\left(\frac{x_i - \bar{X}}{s_x}\right)\left(\frac{y_i - \bar{Y}}{s_y}\right)$$

standardized (z) score

$$z = \frac{\bar{X} - \mu}{\sigma / \sqrt{n}}$$

variance (population)

$$\sigma^2 = \frac{\Sigma(\mu - x_i)^2}{n}$$

variance (sample)

$$s^2 = \frac{\Sigma(\bar{X} - x_i)^2}{n-1}$$

binomial coefficient

$$\binom{n}{k} = \frac{n!}{k!(n-k)!}$$

standard deviation (population)

$$\sigma = \sqrt{\frac{\Sigma(\mu - x_i)^2}{n}}$$

standard deviation (sample)

$$s = \sqrt{\frac{\Sigma(\bar{X} - x_i)^2}{n-1}}$$

binomial probability

$$P(X = k) = \binom{n}{k} p^k (1-p)^{n-k}$$

$$\hat{y} = a + bx$$

$$a = \bar{y} - b\bar{x}$$

$$b = r(sy/sx)$$

least-squares linear regression formula

binomial mean

$$\mu = np$$

$$\sigma = \sqrt{np(1-p)}$$

binomial standard deviation

Statistical formulas

test statistics

standardized (z) score

$$z = \frac{\bar{X} - \mu}{\sigma / \sqrt{n}}$$

one-sample t test

$$t = \frac{\bar{X} - \mu}{\sigma / \sqrt{n}}$$

two-sample t test

$$t = \frac{\bar{X}_1 - \bar{X}_2}{\sqrt{\frac{s_1^2}{n_1} + \frac{s_2^2}{n_2}}}$$

confidence intervals

standardized (z) score

$$CI = \bar{X} \pm z^* \frac{\sigma}{\sqrt{n}}$$

one-sample t test

$$\bar{X} \pm t^* \frac{s}{\sqrt{n}}$$

two-sample t test

$$(\bar{X}_1 - \bar{X}_2) \pm t^* \sqrt{\frac{s_1^2}{n_1} + \frac{s_2^2}{n_2}}$$

sample size

$$n = \left(\frac{z^* \sigma}{m}\right)^2$$

Statistical formulas

sample proportions

$$\hat{p} = \frac{X}{n}$$

X: number of successes

n: total number (sample size)

margin of error for a proportion

$$m = \pm z^* \sqrt{\frac{\hat{p}(1-\hat{p})}{n}}$$

(z* is the critical value corresponding to the confidence level C)

standard deviation

$$\sqrt{\frac{p(1-p)}{n}}$$

confidence interval

$$\hat{p} \pm z^* \sqrt{\frac{\hat{p}(1-\hat{p})}{n}}$$

estimate: $\hat{p}$

margin of error: $z^* \sqrt{\frac{\hat{p}(1-\hat{p})}{n}}$

z score

$$z = \frac{\hat{p} - p}{\sqrt{\frac{p(1-p)}{n}}}$$

sample size for proportions

$$n = \left(\frac{z^*}{m}\right)^2 p^*(1 - p^*)$$

Confidence interval for $p_1 - p_2$ is

$$(\hat{p}_1 - \hat{p}_2) \pm z^* \sqrt{\frac{\hat{p}_1(1 - \hat{p}_1)}{n_1} + \frac{\hat{p}_2(1 - \hat{p}_2)}{n_2}}$$

(estimate): $(\hat{p}_1 - \hat{p}_2)$

(margin of error): $z^* \sqrt{\frac{\hat{p}_1(1 - \hat{p}_1)}{n_1} + \frac{\hat{p}_2(1 - \hat{p}_2)}{n_2}}$

Statistical formulas

pooled sample proportions

pooled sample proportion

$$\hat{p} = \frac{X_1 + X_2}{n_1 + n_2}$$

Sample proportions standard error

(when comparing two sample proportions)

$$se = \sqrt{\frac{\hat{p}_1(1 - \hat{p}_1)}{n_1} + \frac{\hat{p}_2(1 - \hat{p}_2)}{n_2}}$$

z score formula

(To test H_0: $p_1 = p_2$)

$$z = \frac{\hat{p}_1 - \hat{p}_2}{\sqrt{\hat{p}(1 - \hat{p})\left(\frac{1}{n_1} + \frac{1}{n_2}\right)}}$$

($\hat{p}$ in these tests is the pooled sample proportion.)

$$\hat{p} = \frac{X_1 + X_2}{n_1 + n_2}$$

Population proportions standard deviation

(when comparing two populations)

$$\sigma = \sqrt{\frac{p_1(1 - p_1)}{n_1} + \frac{p_2(1 - p_2)}{n_2}}$$

Confidence interval for $p_1 - p_2$ is

$$(\hat{p}_1 - \hat{p}_2) \pm z^* \sqrt{\frac{\hat{p}_1(1 - \hat{p}_1)}{n_1} + \frac{\hat{p}_2(1 - \hat{p}_2)}{n_2}}$$

(estimate) (margin of error)

Statistical formulas

chi-square

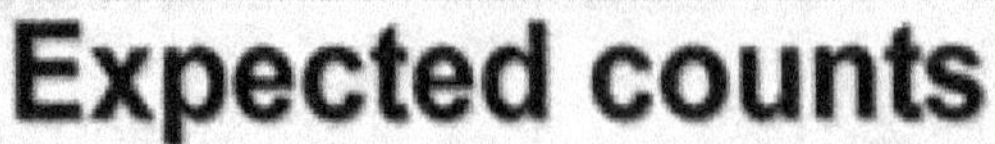

If Ho were true, then the *expected* cell count of a two-way table is...

$$\text{expected cell count} = \frac{\text{row total} \times \text{column total}}{\text{table total}}$$

chi-square

Degrees of freedom

df = (# of rows - 1)(# of columns - 1)

df = (r - 1)(c - 1)

Chi-square test statistic

$$x^2 = \sum \frac{(\text{observed count} - \text{expected count})^2}{\text{expected count}}$$

$$x^2 = \sum \frac{(O - E)^2}{E}$$

(The $\sum$ means to add up all of the cell differences.)

Answers to guided homework (odds only)

All problems from the guided homework sheets are explained and answered via online videos at http://mathguyzero.com/moodle/ .

Chapter 1

1. Categorical: gender, alcohol user, ethnicity and smoker. Quantitative: height, weight, age, income, distance driven daily and blood pressure.

3. 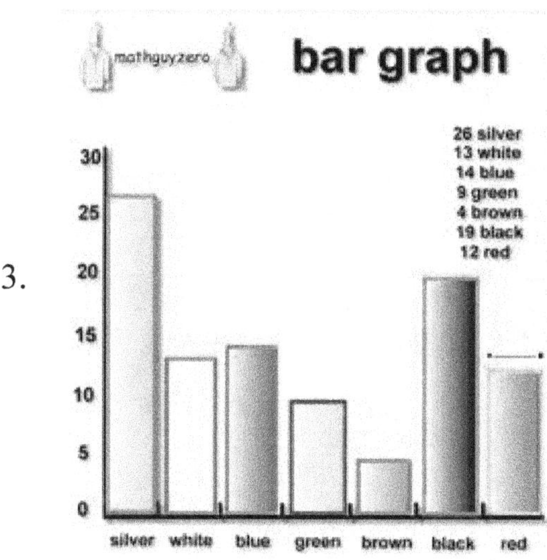

5. 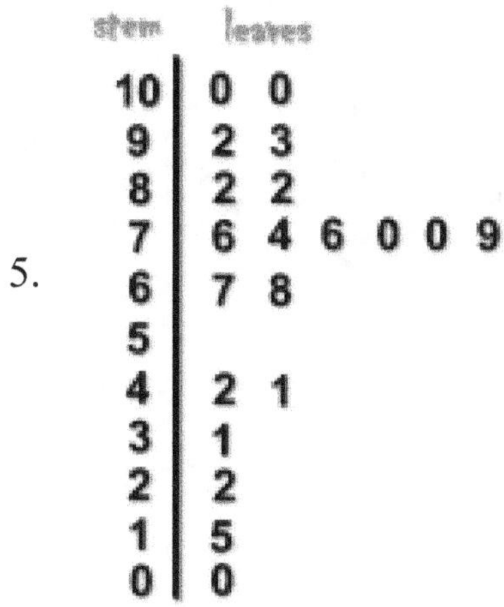

Answers

Guided Homework

Chapter 2

1. Median = 54

3. Mode = 54

5. Standard deviation = 6.3

7.

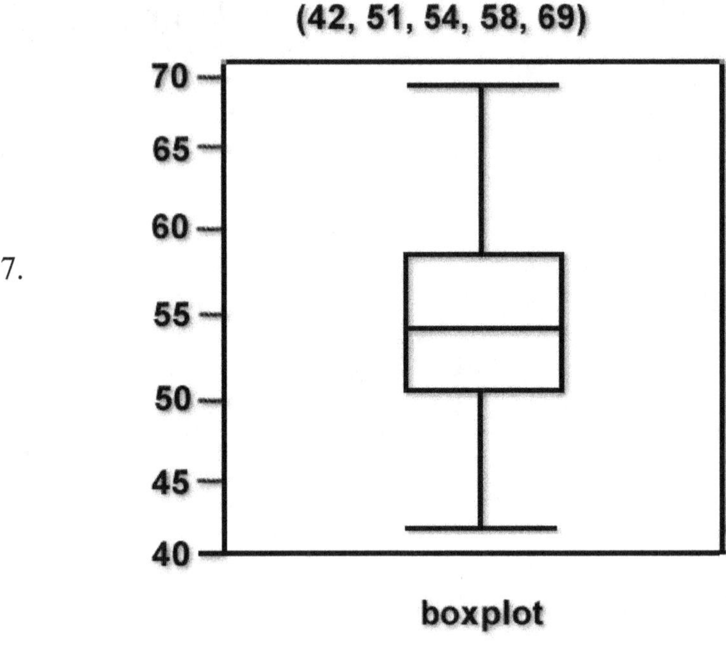

Chapter 3

1. 0.7985

3. 0.2951

5. 0.2920

Answers

Guided Homework

Chapter 4

1. Explanatory variable is the amount of water and the response variable is the height of plant.

3. Strong correlation

5. y = 3.3333x + 205

Chapter 5

1. P(x = red) = 0.18; P(x = grey or white) = 0.43; P(x = not black or white) = 0.70; P(x = purple) $\leq$ 0.02

3. 1/216 or 0.0046

5. 2/9 or 0.2222

7. a) 0.099; b) 0.200; c) 0.400; d) 0.004; e) 0.996

Chapter 6

1. a) Not a binomial setting; b) yes; c) yes

3. a) yes; b) the random variable x is the number of homeowners that lost equity out of the 25 homeowners contacted; c) Possible outcomes: 0, 1, 2, … 24, 25, so SRS = 26 d) n = 25, p = 0.80; e) 0.062 f) x-bar (mean) = 20, standard deviation = 2

5. 1/56 or 0.0179

7. 78 or 79 goals – boo yeah!

Answers

Guided Homework

Chapter 7

1. 0.2839

3. 0.4234

5. 5.3 ft.

7. 0.8289

Chapter 8

1. 103.6 ± .71

3. 7.563 ± .106

5. 612

Chapter 9

1. a) H0: $\mu = 1.672$; b) Ha: $\mu < 1.672$; c) -1.72; d) 0.0427 e) significant - reject the null at the .05 level; f) not significant - fail to reject the null at the .01 level

3. not significant - fail to reject the null at the .05 level

5. not significant - fail to reject the null at the .05 level

Chapter 10

1. standard error = 2.03

3. a) df = 24; b) one-tailed test: $.10 < P(1.02) < .25$, two-tailed test: $.20 < P(1.02) < .50$; c) not significant - fail to reject the null at the .10 level; d) not significant - fail to reject the null at the .05 level

5. significant - reject the null at the .05 level

Answers

Guided Homework

Chapter 11

1. .a) 12 x 72 ≈ 9, b) .12 x 72 = 8.64, c) 0.038, d) 8.64 ± .075 e) 390 f) cultural - left-handed is bad; whichever hand closest to the mouth in utero

3. a) HO: $p_0 = .58$, Ha: $p_0 \neq 5.8$; b) z score = 0.81, p-value = 0.418, do not reject null,

5. a) H0: $P_m = P_f$ and Ha: $P_m > P_f$, b) . p-hat males is 25/35 ≈ .71, p-hat females is 42/63 ≈ .67 , c) .0970, d) 04 ± .25 or [-.21, .29] Warning!! Any CI that contains zero is reason to NOT reject the null!
e) around 69%, f) z score = 0.42, p-value = 0.6591, do not reject null

Chapter 12

1. H0: $p_i = .725.8$ Ha: $p_i \neq .725.8$ $X^2 = 2.94$, df = 4, p-value > 0.25, do not reject the null – no significant difference

3. H0: no relationship between music genre and mood Ha: there is a relationship – one categorical variable is dependent on the other categorical variable, $X^2 = 16.62$, df = 2, p-value < 0.0005, reject the null – significant relationship

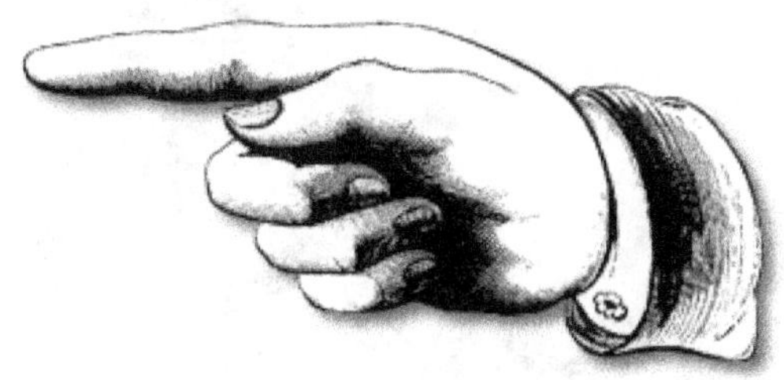

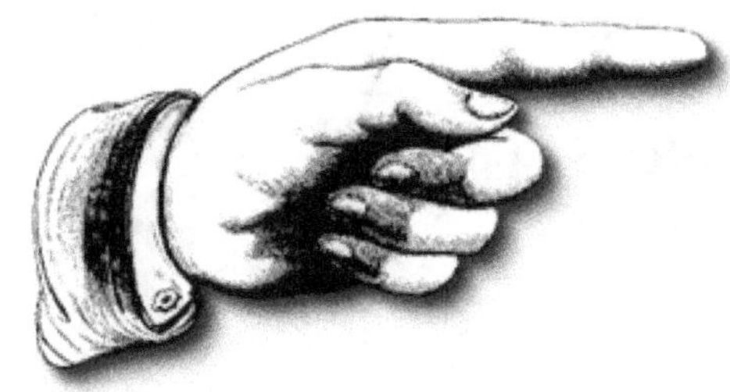

Answers to quizzes, midterm and final exam (odds only)

Answers

Chapter Quizzes

Quiz: Chapter 1

1. subjects

3. false

5. Friday

7. d

9. a

Quiz: Chapter 2

1. mean = 66.14

3. mode = 72

5. standard deviation = 24.51

7. q3 = 81

9. make the boxplot yourself!

11. around 200

Quiz: Chapter 3

1. 33.3%

3. .0146

5. 0

7. true

9. a) around 16%; b) between 3.61 and 4.93 oz; c) 3.61 oz

Answers

Chapter Quizzes

Quiz: Chapter 4

1. a) Explanatory variable is the amount of the drug dosage and the response variable is the difference between pre and post test scores; b) r = .61; c) the correlation is weak (or not very strong) and positive; d) make your own scatterplot; e) y = 11.43x + 3.52; f) around 13.24

3. a

5. d

Quiz: Chapter 5

1. $6^5 = 7776$

3. a) 0.15; b) 2.52%; c) $P(x = \text{Yankees}) \leq 0.13$; d) 0.67; e) 0.39

5. a) 0.50; b) around 2%

7. 1/6 just like always!

9. around .000000003

Answers

Midterm

Midterm: Chapters 1 - 5

1: d

3a: around 2%, 3b: 50%

5: e

7a: 0.98, 7b: c, 7c: y = 20.6x + 1, 7d: around 16

9a: 3.45%, 9b: less than or equal to 13%, 9c: 27%, 9d: around 61%

11: 2.08

13a: around 44%, 13b: 140

15a: around $31,500, 15b: 75%

17: around 42%

19: c

21: 1/6, same as every roll

23: 3/52 or around 5.8%

25: 1

27: 0.9421

29: d

Quiz: Chapter 6

1. c

3. 330

5. around 0.83%

7. 1/15504 (That is if he wasn't truly psychic!)

9. a

11. true

Quiz: Chapter 7

1. a) around 28%; b) around 53%; c) around 2%; d) around 29%; e) around 50%

3. 3

5. b

7. 100

Quiz: Chapter 8

1. 29.19 ± .54 grams

3. d

5. 1439

7. false

9. false – we need the SAMPLE mean!

11. 55 ± 1.55 grams

13. 69 ± 0.52

15. ± 3.09

17. true

19. a

Quiz: Chapter 9

1. b

3. true

5. true

7. 0.0244

9. false

11. 0.35

13. false

15. false

17. d

19. true

21. true

Quiz: Chapter 10

1. true

3. between 10% and 20%

5. true

7. a) -3.83 ± 4.06; b) H0: $\mu = 0$ and Ha: $\mu \neq 0$; c) $|-2.43| = 2.43$; d) between .05 and .10; e) not significant at the .05 level – do not reject the null; f) not significant at the .05 level – do not reject the null

9. false: df = n – 1

Quiz: Chapter 11

1. a) 56/125 =0 .448, b) 0.39 x 125 = 48.75 , c) 0.044,
d) 48.75 ± 0.086, e) 632

3. a) H0: p = .55, Ha: p > .55 , b) p-hat = .5567, z score = 0.42, p-value =0 .3372 … therefore this does support the stated population proportion – do not reject the null.

5. males p1, females p2, p-hat1 ≈ .58 and p-hat2 ≈ .67, pooled p-hat ≈ .62, pooled standard error ≈ .05, z score = 1.8, p-value = .07, do not reject the null

Answers

Chapter Quizzes

Quiz: Chapter 12

1. H0: $p_i = 29.4$, Ha: $p_i \neq 29.4$, $X^2 = 10.993$, df = 4, p-value = 0 .027 reject the null - significant difference

3. H0: there is no relationship between the categorical variables, Ha: there is a relationship, $X^2 = 2.11$, df = 4, p-value = 0.35, do not reject the null no significant relationship

Final Exam: Chapters 6 - 12

Answers

Final Exam

1a: 800,1b: 12.65, 1c: around 7%

3a: 0.36 inches

5: true

7a: e

9a: around 0.7%, 9b: around 2%, 9c: around 66%, 9d: around 4%, 9e: around 50%

11: b

13: b

15: true

17: c

19: true

21: a

Answers

Final Exam

23: around 0.44%

25: c

27: 165.8 +/- 9.08

29: 1/575757 or around 0.00017%

Answers

Final Exam

31: true

33: +/- 3.55

35: 1365 - that's a lot of pizzas!

37: true

39: false

41: c

43: false

45: e

47. a

49. e

t distribution table

t distribution critical values table

d.f.	Level of confidence, c	0.50	0.80	0.90	0.95	0.98	0.99
	One tail, α	0.25	0.10	0.05	0.025	0.01	0.005
	Two tails, α	0.50	0.20	0.10	0.05	0.02	0.01
1		1.000	3.078	6.314	12.706	31.821	63.657
2		.816	1.886	2.920	4.303	6.965	9.925
3		.765	1.638	2.353	3.182	4.541	5.841
4		.741	1.533	2.132	2.776	3.747	4.604
5		.727	1.476	2.015	2.571	3.365	4.032
6		.718	1.440	1.943	2.447	3.143	3.707
7		.711	1.415	1.895	2.365	2.998	3.499
8		.706	1.397	1.860	2.306	2.896	3.355
9		.703	1.383	1.833	2.262	2.821	3.250
10		.700	1.372	1.812	2.228	2.764	3.169
11		.697	1.363	1.796	2.201	2.718	3.106
12		.695	1.356	1.782	2.179	2.681	3.055
13		.694	1.350	1.771	2.160	2.650	3.012
14		.692	1.345	1.761	2.145	2.624	2.977
15		.691	1.341	1.753	2.131	2.602	2.947
16		.690	1.337	1.746	2.120	2.583	2.921
17		.689	1.333	1.740	2.110	2.567	2.898
18		.688	1.330	1.734	2.101	2.552	2.878
19		.688	1.328	1.729	2.093	2.539	2.861
20		.687	1.325	1.725	2.086	2.528	2.845
21		.686	1.323	1.721	2.080	2.518	2.831
22		.686	1.321	1.717	2.074	2.508	2.819
23		.685	1.319	1.714	2.069	2.500	2.807
24		.685	1.318	1.711	2.064	2.492	2.797
25		.684	1.316	1.708	2.060	2.485	2.787
26		.684	1.315	1.706	2.056	2.479	2.779
27		.684	1.314	1.703	2.052	2.473	2.771
28		.683	1.313	1.701	2.048	2.467	2.763
29		.683	1.311	1.699	2.045	2.462	2.756
∞		.674	1.282	1.645	1.960	2.326	2.576

Standardized normal distribution table (z table)

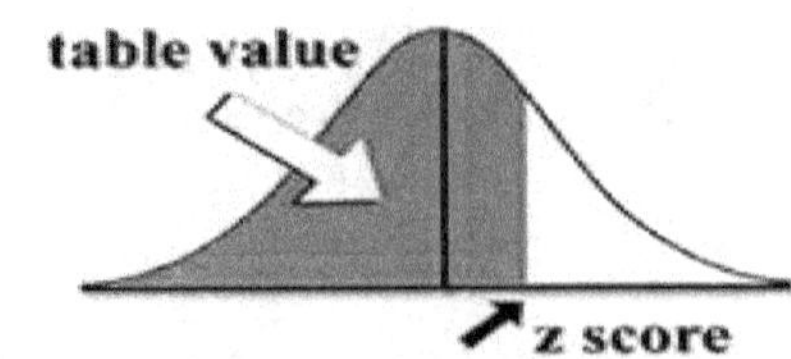

z score
(positive)

Standard Normal Probabilities

	0	0.01	0.02	0.03	0.04	0.05	0.06	0.07	0.08	0.09
0	.5000	.5040	.5080	.5120	.5160	.5199	.5239	.5279	.5319	.5359
0.1	.5398	.5438	.5478	.5517	.5557	.5596	.5636	.5675	.5714	.5753
0.2	.5793	.5832	.5871	.5910	.5948	.5987	.6026	.6064	.6103	.6141
0.3	.6179	.6217	.6255	.6293	.6331	.6368	.6406	.6443	.6480	.6517
0.4	.6554	.6591	.6628	.6664	.6700	.6736	.6772	.6808	.6844	.6879
0.5	.6915	.6950	.6985	.7019	.7054	.7088	.7123	.7157	.7190	.7224
0.6	.7257	.7291	.7324	.7357	.7389	.7422	.7454	.7486	.7517	.7549
0.7	.7580	.7611	.7642	.7673	.7704	.7734	.7764	.7794	.7823	.7852
0.8	.7881	.7910	.7939	.7967	.7995	.8023	.8051	.8078	.8106	.8133
0.9	.8159	.8186	.8212	.8238	.8264	.8289	.8315	.8340	.8365	.8389
1	.8413	.8438	.8461	.8485	.8508	.8531	.8554	.8577	.8599	.8621
1.1	.8643	.8665	.8686	.8708	.8729	.8749	.8770	.8790	.8810	.8830
1.2	.8849	.8869	.8888	.8907	.8925	.8944	.8962	.8980	.8997	.9015
1.3	.9032	.9049	.9066	.9082	.9099	.9115	.9131	.9147	.9162	.9177
1.4	.9192	.9207	.9222	.9236	.9251	.9265	.9279	.9292	.9306	.9319
1.5	.9332	.9345	.9357	.9370	.9382	.9394	.9406	.9418	.9429	.9441
1.6	.9452	.9463	.9474	.9484	.9495	.9505	.9515	.9525	.9535	.9545
1.7	.9554	.9564	.9573	.9582	.9591	.9599	.9608	.9616	.9625	.9633
1.8	.9641	.9649	.9656	.9664	.9671	.9678	.9686	.9693	.9699	.9706
1.9	.9975	.9975	.9726	.9732	.9738	.9744	.9750	.9756	.9761	.9767
2	.9772	.9778	.9783	.9788	.9793	.9798	.9803	.9808	.9812	.9817
2.1	.9821	.9826	.9830	.9834	.9838	.9842	.9846	.9850	.9854	.9857
2.2	.9861	.9864	.9868	.9871	.9875	.9878	.9881	.9884	.9887	.9890
2.3	.9893	.9896	.9898	.9901	.9904	.9906	.9909	.9911	.9913	.9916
2.4	.9918	.9920	.9922	.9925	.9927	.9929	.9931	.9932	.9934	.9936
2.5	.9938	.9940	.9941	.9943	.9945	.9946	.9948	.9949	.9951	.9952
2.6	.9953	.9955	.9956	.9957	.9959	.9960	.9961	.9962	.9963	.9964
2.7	.9965	.9966	.9967	.9968	.9969	.9970	.9971	.9972	.9973	.9974
2.8	.9974	.9975	.9976	.9977	.9977	.9978	.9979	.9979	.9980	.9981
2.9	.9981	.9982	.9982	.9983	.9984	.9984	.9985	.9985	.9986	.9986
3	.9987	.9987	.9987	.9988	.9988	.9989	.9989	.9989	.9990	.9990
3.1	.9990	.9991	.9991	.9991	.9992	.9992	.9992	.9992	.9993	.9993
3.2	.9993	.9993	.9994	.9994	.9994	.9994	.9994	.9995	.9995	.9995
3.3	.9995	.9995	.9995	.9996	.9996	.9996	.9996	.9996	.9996	.9997
3.4	.9997	.9997	.9997	.9997	.9997	.9997	.9997	.9997	.9997	.9998

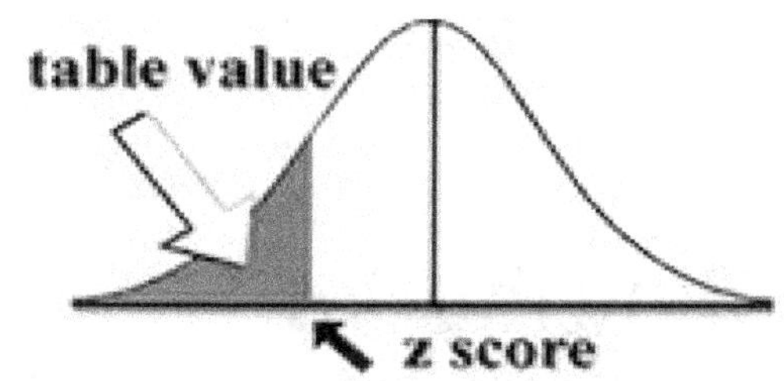

z score
(negative)

Standard Normal Probabilities

	0	0.01	0.02	0.03	0.04	0.05	0.06	0.07	0.08	0.09
0	.5000	.4960	.4920	.4880	.4840	.4801	.4761	.4721	.4681	.4641
-0.1	.4602	.4562	.4522	.4483	.4443	.4404	.4364	.4325	.4286	.4247
-0.2	.4207	.4168	.4129	.4090	.4052	.4013	.3974	.3936	.3897	.3859
-0.3	.3821	.3783	.3745	.3707	.3669	.3632	.3594	.3557	.3520	.3483
-0.4	.3446	.3409	.3372	.3336	.3300	.3264	.3228	.3192	.3156	.3121
-0.5	.3085	.3050	.3015	.2981	.2946	.2912	.2877	.2843	.2810	.2776
-0.6	.2743	.2709	.2676	.2643	.2611	.2578	.2546	.2514	.2483	.2451
-0.7	.2420	.2389	.2358	.2327	.2296	.2266	.2236	.2206	.2177	.2148
-0.8	.2119	.2090	.2061	.2033	.2005	.1977	.1949	.1922	.1894	.1867
-0.9	.1841	.1814	.1788	.1762	.1736	.1711	.1685	.1660	.1635	.1611
-1	.1587	.1562	.1539	.1515	.1492	.1469	.1446	.1423	.1401	.1379
-1.1	.1357	.1335	.1314	.1292	.1271	.1251	.1230	.1210	.1190	.1170
-1.2	.1151	.1131	.1112	.1093	.1075	.1056	.1038	.1020	.1003	.0985
-1.3	.0968	.0951	.0934	.0918	.0901	.0885	.0869	.0853	.0838	.0823
-1.4	.0808	.0793	.0778	.0764	.0749	.0735	.0721	.0708	.0694	.0681
-1.5	.0668	.0655	.0643	.0630	.0618	.0606	.0594	.0582	.0571	.0559
-1.6	.0548	.0537	.0526	.0516	.0505	.0495	.0485	.0475	.0465	.0455
-1.7	.0446	.0436	.0427	.0418	.0409	.0401	.0392	.0384	.0375	.0367
-1.8	.0359	.0351	.0344	.0336	.0329	.0322	.0314	.0307	.0301	.0294
-1.9	.0287	.0281	.0274	.0268	.0262	.0256	.0250	.0244	.0239	.0233
-2	.0228	.0222	.0217	.0212	.0207	.0202	.0197	.0192	.0188	.0183
-2.1	.0179	.0174	.0170	.0166	.0162	.0158	.0154	.0150	.0146	.0143
-2.2	.0139	.0136	.0132	.0129	.0125	.0122	.0119	.0116	.0113	.0110
-2.3	.0107	.0104	.0102	.0099	.0096	.0094	.0091	.0089	.0087	.0084
-2.4	.0082	.0080	.0078	.0075	.0073	.0071	.0069	.0068	.0066	.0064
-2.5	.0062	.0060	.0059	.0057	.0055	.0054	.0052	.0051	.0049	.0048
-2.6	.0047	.0045	.0044	.0043	.0041	.0040	.0039	.0038	.0037	.0036
-2.7	.0035	.0034	.0033	.0032	.0031	.0030	.0029	.0028	.0027	.0026
-2.8	.0026	.0025	.0024	.0023	.0023	.0022	.0021	.0021	.0020	.0019
-2.9	.0019	.0018	.0018	.0017	.0016	.0016	.0015	.0015	.0014	.0014
-3	.0013	.0013	.0013	.0012	.0012	.0011	.0011	.0011	.0010	.0010
-3.1	.0010	.0009	.0009	.0009	.0008	.0008	.0008	.0008	.0007	.0007
-3.2	.0007	.0007	.0006	.0006	.0006	.0006	.0006	.0005	.0005	.0005
-3.3	.0005	.0005	.0005	.0004	.0004	.0004	.0004	.0004	.0004	.0003
-3.4	.0003	.0003	.0003	.0003	.0003	.0003	.0003	.0003	.0003	.0002

Chi-square distribution table

Chi-square (X^2) distribution critical values

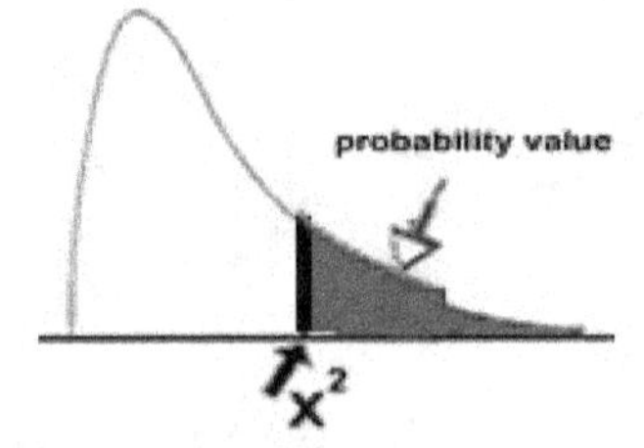

probability values

df	0.25	0.20	0.15	0.10	0.05	0.025	0.020	0.010	0.005	0.0025	0.001	0.0005
1	1.32	1.64	2.07	2.71	3.84	5.02	5.41	6.63	7.88	9.14	10.83	12.12
2	2.77	3.22	3.79	4.61	5.99	7.38	7.82	9.21	10.60	11.98	13.82	15.20
3	4.11	4.64	5.32	6.25	7.81	9.35	9.84	11.34	12.84	14.32	16.27	17.73
4	5.39	5.99	6.74	7.78	9.49	11.14	11.67	13.28	14.86	16.42	18.47	20.00
5	6.63	7.29	8.12	9.24	11.07	12.83	13.39	15.09	16.75	18.39	20.52	22.11
6	7.84	8.56	9.45	10.64	12.59	14.45	15.03	16.81	18.55	20.25	22.46	24.10
7	9.04	9.80	10.75	12.02	14.07	16.01	16.62	18.48	20.28	22.04	24.32	26.02
8	10.22	11.03	12.03	13.36	15.51	17.53	18.17	20.09	21.95	23.77	26.12	27.87
9	11.39	12.24	13.29	14.68	16.92	19.02	19.68	21.67	23.59	25.46	27.88	29.67
10	12.55	13.44	14.53	15.99	18.31	20.48	21.16	23.21	25.19	27.11	29.59	31.42
11	13.70	14.63	15.77	17.28	19.68	21.92	22.62	24.72	26.76	28.73	31.26	33.14
12	14.85	15.81	16.99	18.55	21.03	23.34	24.05	26.22	28.30	30.32	32.91	34.82
13	15.98	16.98	18.20	19.81	22.36	24.74	25.47	27.69	29.82	31.88	34.53	36.48
14	17.12	18.15	19.41	21.06	23.68	26.12	26.87	29.14	31.32	33.43	36.12	38.11
15	18.25	19.31	20.60	22.31	25.00	27.49	28.26	30.58	32.80	34.95	37.70	39.72
16	19.37	20.47	21.79	23.54	26.30	28.85	29.63	32.00	34.27	36.46	39.25	41.31
17	20.49	21.61	22.98	24.77	27.59	30.19	31.00	33.41	35.72	37.95	40.79	42.88
18	21.60	22.76	24.16	25.99	28.87	31.53	32.35	34.81	37.16	39.42	42.31	44.43
19	22.72	23.90	25.33	27.20	30.14	32.85	33.69	36.19	38.58	40.88	43.82	45.97
20	23.83	25.04	26.50	28.41	31.41	34.17	35.02	37.57	40.00	42.34	45.31	47.50
21	24.93	26.17	27.66	29.62	32.67	35.48	36.34	38.93	41.40	43.78	46.80	49.01
22	26.04	27.30	28.82	30.81	33.92	36.78	37.66	40.29	42.80	45.20	48.27	50.51
23	27.14	28.43	29.98	32.01	35.17	38.08	38.97	41.64	44.18	46.62	49.73	52.00
24	28.24	29.55	31.13	33.20	36.42	39.36	40.27	42.98	45.56	48.03	51.18	53.48
25	29.34	30.68	32.28	34.38	37.65	40.65	41.57	44.31	46.93	49.44	52.62	54.95
26	30.43	31.79	33.43	35.56	38.89	41.92	42.86	45.64	48.29	50.83	54.05	56.41
27	31.53	32.91	34.57	36.74	40.11	43.19	44.14	46.96	49.64	52.22	55.48	57.86
28	32.62	34.03	35.71	37.92	41.34	44.46	45.42	48.28	50.99	53.59	56.89	59.30
29	33.71	35.14	36.85	39.09	42.56	45.72	46.69	49.59	52.34	54.97	58.30	60.73
30	34.80	36.25	37.99	40.26	43.77	46.98	47.96	50.89	53.67	56.33	59.70	62.16
40	45.62	47.27	49.24	51.81	55.76	59.34	60.44	63.69	66.77	69.70	73.40	76.09
50	56.33	58.16	60.35	63.17	67.50	71.42	72.61	76.15	79.49	82.66	86.66	89.56
60	66.98	68.97	71.34	74.40	79.08	83.30	84.58	88.38	91.95	95.34	99.61	102.69
80	88.13	90.41	93.11	96.58	101.88	106.63	108.07	112.33	116.32	120.10	124.84	128.26
100	109.14	111.67	114.66	118.50	124.34	129.56	131.14	135.81	140.17	144.29	149.45	153.17

www.ingramcontent.com/pod-product-compliance
Lightning Source LLC
Chambersburg PA
CBHW081110170526
45165CB00008B/2391

9781463731342